Lecture Notes in Economics and Mathematical Systems

605

Michael Puhle

Bond Portfolio Optimization

Springer

Dr. Michael Puhle
Allianz Global Investors Kapitalanlagegesellschaft mbH
Nymphenburger Straße 112-116
80636 Munich
Germany
michael.puhle@allianzgi.de

Doctoral thesis, University of Passau, 2007

ISBN 978-3-540-76592-9 e-ISBN 978-3-540-76593-6

DOI 10.1007/978-3-540-76593-6

Lecture Notes in Economics and Mathematical Systems ISSN 0075-8442

Library of Congress Control Number: 2007938963

Production: LE-TEX Jelonek, Schmidt & Vöckler GbR, Leipzig
Cover design: WMX Design GmbH, Heidelberg

9 8 7 6 5 4 3 2 1

springer.com

Acknowledgements

I would like to express my gratitude to a number of people. First of all, I thank my thesis advisor Prof. Dr. Jochen Wilhelm for supervising and guiding me during my years at the chair of finance at Passau University. Prof. Dr. János Száz gave me the opportunity to present previous versions at seminars in Budapest. He also provided valuable feedback and agreed to be the second referee.

Prof. Dr. Bernhard Nietert took the time to discuss several parts of this thesis with me. Dipl.-Kffr. Marion Trautbeck-Kim and Dipl.-Kfm. Andreas Kremer read a draft of this thesis on short notice. Furthermore, I would like to thank Dr. Wolfram Peters for giving me the opportunity to complete this thesis during my first months at Allianz Global Investors.

My parents and my brother supported this project from the beginning and provided warm encouragement during difficult days. Last but not least, I would like to thank my fiancée Veronika for her support especially during the final stages.

Contents

Abbreviations

CIR	Cox/Ingersoll/Ross (1985)
CRRA	Constant relative risk aversion
EUR	Currency code for the Euro
HJB	Hamilton-Jacobi-Bellman equation
HJM	Heath/Jarrow/Morton (1992)
HW2	Hull/White (1994)
InvG	Investmentgesetz (German Investment Act)
MA	Martingale approach
MMA	Money market account
ODE	Ordinary differential equation
PDE	Partial differential equation
RRA	Relative risk aversion
SCA	Stochastic control approach
SDE	Stochastic differential equation
USD	Currency code US Dollar
e.g.	Exempli gratia, for example
ff.	And following pages
i.e.	Id est, that is
max	Maximize
min	Minimize
p.	Page
pp.	Pages
s.t.	Subject to
std.	Standard deviation
w.r.t.	With respect to

Commonly Used Symbols

t	Time
T	Maturity date or investment horizon
$R(t,T)$	Continuously-compounded spot interest rate from t to T
$P(t,T)$	Price at time t for a zero-coupon bond with maturity date T
μ	Drift of the zero-coupon bond price
σ_i	i-th volatility of the zero-coupon bond price, $1 \leq i \leq d$
λ_i	i-th market price of interest rate risk, $1 \leq i \leq d$
$r(t)$	Short rate of interest, $r(t) = f(t,t)$
α	Drift of the short rate
σ_r	Volatility of the short rate
${}_tR(T,\tau)$	Forward interest rate set at time t for a loan that starts at time T and is to be repaid at time τ
$f(t,T)$	Instantaneous forward rate set at time t for a loan that begins at time T and is to be repaid an instant later
m	Drift of the instantaneous forward rates
s_i	i-th volatility of the instantaneous forward rates, $1 \leq i \leq d$
dt	Instantaneous time period
Δt	Short time period
dz	Vector of Brownian motion increments ($d \times 1$)
d	Number of Brownian motions
$B(t)$	Value at time t of a money market account with $B(0) = 1$
$E[x]$	Expectation of x
std(x)	Standard deviation of x
var(x)	Variance of x
cov(x,y)	Covariance between x and y
corr(x,y)	Correlation between x and y
ζ	Stochastic discount factor
a	Drift of the stochastic discount factor
b_i	i-th volatility of the stochastic discount factor, $1 \leq i \leq d$
N	Holdings vector

$\hat{N}$	Holdings vector of risky zero-coupon bonds
W_t	Wealth at time t
n	Number of assets
τ	Maximum maturity of zero-coupon bonds
$\hat{P}_0$	Vector of current prices of risky zero-coupon bonds
C	Covariance matrix
ε	Second factor in HW2 model
ρ	Correlation between r and ε in HW2 model
θ	Mean reversion level
κ_r	Mean reversion speed of short rate in Vasicek and HW2 model
κ_ε	Mean reversion speed of ε in HW2 model
σ_ε	Volatility of ε in HW2 model
$u(W)$	Utility of wealth function
γ	Risk aversion parameter in CRRA utility function
x	Vector of state variables $(d \times 1)$
d	Number of state variables
α	Drift vector of state variables
β	Volatility matrix of state variables
σ_t	Matrix of volatilities of bond price returns $(n \times d)$
w	Portfolio weights vector
w^*	Optimum portfolio weights vector
J	Optimal value function
J_x	Partial derivative of J with respect to x
$\underline{I}$	Identity matrix

1

Introduction

The tools of modern portfolio theory[1] are in general use in the equity markets, either in the form of portfolio optimization software or as an accepted framework in which the asset managers think about stock selection.[2] In the fixed income market on the other hand, these tools seem irrelevant or inapplicable. Bond portfolios are nowadays mainly managed by a comparison of portfolio risk measures[3] vis á vis a benchmark.[4] The portfolio manager's views about the future evolution of the term structure of interest rates translate themselves directly into a positioning relative to his benchmark, taking the risks of these deviations from the benchmark into account only in a very crude fashion, i.e. without really quantifying them probabilistically.[5] This is quite surprising since sophisticated models for the evolution of interest rates are commonly used for interest rate derivatives pricing and the derivation of fixed income risk measures.[6]

Wilhelm (1992) explains the absence of modern portfolio tools in the fixed income markets with two factors:[7] historically relatively stable interest rates and systematic differences between stocks and bonds that make an application of modern portfolio theory difficult. These systematic differences relate mainly to the fixed maturity of bonds. Whereas possible future stock prices become more dispersed as the time horizon widens, the bond price at maturity is fixed.[8] This implies that the probabilistic models for stocks and bonds have

---

1 Starting with the seminal work of Markowitz (1952).

2 See e.g. Grinold/Kahn (2000), Litterman (2003) or Elton et al. (2003).

3 These are commonly partial duration like risk measures, such as level, slope and curvature durations introduced in Willner (1996).

4 See the standard literature on bond portfolio selection, e.g. Fabozzi (2001).

5 Usually the tracking error is calculated but doesn't receive much attention.

6 For example to derive a duration measure for swaptions. For a review of commonly used bond portfolio risk measures see Golub/Tilman (2000).

7 Wilhelm (1992), p. 210.

8 It must be equal to the face value plus coupon. The fixed income terminology will be presented in Chapter 2.

to be different and the tools of modern portfolio theory have to be adapted to be applicable to fixed income instruments.[9]

In this thesis, we analyze how modern portfolio theory[10] and dynamic term structure models can be used for government bond portfolio optimization. We study the necessary adjustments, examine the models with regard to the plausibility of their results and compare the outcomes to portfolio selection techniques used by practitioners.

The question of how to adapt the mean-variance model for bond portfolio selection purposes is hardly new. The earliest attempt is due to Cheng (1962). He modeled the trade-off between rolling over short-term investments and investing at a given spot rate until the investment horizon.[11] Hence, he confined his analysis to the impact of reinvestment risks on bond portfolio selection.[12] The model needs as input probability beliefs about future term structures of interest rates (as reinvestment rates). Cheng (1962) suggests analyzing empirical data on interest rate movements for a formulation of these probability beliefs.[13] He allows for all possible reinvestment tactics until the investment horizon.[14] One problem is therefore the enormous number of possible tactics to be considered.[15]

Bradley/Crane (1972) propose a dynamic bond portfolio selection formulation. They argue that the approach by Cheng (1962) doesn't take into consideration the dynamic nature of any portfolio selection problem. But as they admit, formulation of a realistic bond portfolio selection problem in their framework results in extensive data requirements.[16] This stems from the fact that one needs to specify an event sequence (possible term structure movements) over time with the associated probability beliefs.[17] With hindsight, the usage of dynamic term structure models for the generation of these probability beliefs seems obvious. But the first dynamic term structure model was only introduced in 1977 by Vasicek.[18] In 1980, Brennan and Schwartz expressed the hope, that using term structure models for bond portfolio selection purposes, will help to reduce the data requirement burden.[19]

---

[9] Hence the geometric Brownian motion model used to model stocks can't be used for bond price modeling purposes.

[10] The static mean-variance model and dynamic continuous-time variants.

[11] See Cheng (1962), p. 492.

[12] He ruled out selling bonds before maturity, so market risks or practical strategies such as rolling down the yield curve (to be introduced in Chapter 4) are ignored.

[13] See Cheng (1962), p. 491.

[14] If for example the investment horizon is 10 years, then one possible tactic is investing for one period at spot interest rate $R(0,1)$, then for four periods at $R(1,5)$, then for another two periods at $R(5,7)$ and at last for three periods at $R(7,10)$ until the investment horizon.

[15] See Cheng (1962), p. 499.

[16] See Bradley/Crane (1972), p. 150.

[17] See Bradley/Crane (1972), p. 150.

[18] Vasicek (1977).

Wilhelm (1992) introduces a bond portfolio selection model using dynamic term structure models. He identifies the portfolio selection problem as a dynamic problem but then comes to the conclusion that the problem in such generality is practically unsolvable and hence he confines his analysis to a static framework.[20] He then derives optimum portfolios in a mean-variance framework where the bond market is governed by the term structure model of Cox/Ingersoll/Ross (1985). A critical assumption of the model is the reinvestment of cash flows occurring before the investment horizon at the current spot interest rate until the investment horizon.[21]

Fabozzi/Fong (1994) describe the possibility of using portfolio optimization in the fixed income markets and identify the calculation of the covariance matrix of bond returns as the main obstacle.[22] They conclude that "if a covariance matrix could be created, the bond optimization process could parallel the analysis for stocks".[23]

Elton et al. (2003) also propagate using modern portfolio theory for bond portfolio selection purposes.[24] They assume implicitly a short-term investment horizon, since they don't make any assumption regarding the reinvestment of cash flows. Furthermore, they use separate models for deriving the probability beliefs about expected returns and covariances of returns. For estimation of expected returns, they suggest using any of the classical term structure theories[25], e.g. the expectation theory.[26] The covariance matrix shall be estimated using a single- or multi-index model.[27] Hence, according to this approach, each bond is influenced by market risk factors and an idiosyncratic risk factor.[28]

A recent contribution to this line of research is due to Korn/Koziol (2006). They analyze the problem of an investor who can invest in zero-coupon bonds of different maturities. The underlying bond market uncertainty is driven by an (multi-factor) affine term structure model.[29] The key contribution of their paper is an examination of the historical performance of mean-variance efficient bond portfolios in the German government bond market.

Another line of research analyzes the adaption of continuous-time portfolio selection model based on Merton (1971) to the selection of bond portfolios.

---

[19] See Brennan/Schwartz (1980), p. 406.

[20] See Wilhelm (1992), p. 216.

[21] See Wilhelm (1992), p. 216.

[22] See Fabozzi/Fong (1994), p. 154.

[23] Fabozzi/Fong (1994), p. 154.

[24] See Elton et al. (2003), pp. 540–546.

[25] To be introduced in Chapter 2.5.

[26] See Elton et al. (2003), p. 540.

[27] See Elton et al. (2003), pp. 543–546.

[28] For default-free government bonds, this seems problematic. All government bond prices (at least of a specific country) are influenced by the same market factors, i.e. interest rates of different maturities. In our opinion there is no room for idiosyncratic risk factors.

[29] See Korn/Koziol (2006), p. 4.

Sørensen (1999) examines the problem of an investor with constant relative risk aversion (CRRA) who can invest into a stock index, a zero-coupon bond and a money market account.[30] The term structure of interest rates is governed by the Vasicek (1977) model. He derives a solution using the martingale approach by Cox/Huang (1989).

Korn/Kraft (2002) study a pure bond portfolio selection problem where interest rates follow either the Vasicek (1977) or the Cox/Ingersoll/Ross (1985) term structure model. They solve the problem using the stochastic control approach. Kraft (2004) extends this analysis by covering additionally the term structure models by Dothan (1978), Ho/Lee (1986) and Black/Karasinksi (1991).

Munk/Sørensen (2004) solve a pure bond portfolio selection problem in a general Heath/Jarrow/Morton (1992) term structure framework using the martingale approach.

In this thesis we derive the Heath/Jarrow/Morton (1992) term structure framework and two special cases (the Vasicek (1977) and the Hull/White (1994) models) using the stochastic discount factor pricing methodology. Hence we use a different approach than is usually taken.[31] Furthermore, we extend the mean-variance bond portfolio selection model proposed by Wilhelm (1992) to the term structure models of Vasicek (1977) and Hull/White (1994). We compare the resulting portfolios to traditional active and passive bond portfolio selection methods. In addition, we derive an explicit analytic solution to a continuous-time bond portfolio selection model where the bond market is driven by the Hull/White (1994) model. As an extension to the continuous-time model, we show how foreign currency bonds can be incorporated in the analysis. We introduce an international bond portfolio selection problem and give an analytic solution for a simple two-country case.

This thesis is organized as follows. Chapter 2 introduces interest rate terminology and the term structure of interest rates. In Chapter 3 we present the derivation of the general term structure modeling framework by Heath/ Jarrow/Morton (1992) using the stochastic discount factor methodology. Furthermore, we derive the Vasicek (1977) and the Hull/White (1994) models as special cases. Chapter 4 deals with static mean-variance bond portfolio selection as introduced by Wilhelm (1992). Chapter 5 examines continuous-time bond portfolio selection problems. We derive the solution to a pure bond portfolio selection problem with the Vasicek (1977) and the Hull/White (1994) models using the stochastic control approach. Furthermore, we analyze an international bond portfolio selection problem and derive an explicit solution for a two-country case. Chapter 6 concludes the thesis.

---

[30] See Sørensen (1999), p. 517.

[31] Originally term structure models were derived using the martingale or the PDE approach.

2

Bond Market Terminology

By a *fixed-income market*, we mean that particular sector of the financial market on which interest rate sensitive instruments trade.[32] In this thesis we are only interested in the *government bond market*, i.e. that part of the fixed-income market where government bonds are traded.

2.1 Characteristics of Bonds

A *bond* represents a claim on a prescribed sequence of payments.[33] The issuer of the bond (the borrower) is committed to paying back to the bondholder (the lender) the cash amount borrowed plus periodic interest payments.[34]

One important characteristic of a bond is the nature of its *issuer*.[35] Typical bond issuers are (municipal and federal) governments and (domestic and foreign) corporations. Government bonds are assumed to be default-free, i.e. the above mentioned sequence of payments is ex ante known with certainty.

Another key feature of a bond is the *maturity (date)*. This is the date on which the debt will cease to exist[36] and the borrower will redeem the issue by paying the *face value*.[37] By *term to maturity* we mean the time to maturity. *Face value* (or principal, or par value) is the amount the issuer agrees to repay at maturity[38] or the nominal amount borrowed by the debtor.

Regarding the sequence of payments the bondholder is entitled to, one can distinguish between two main types of bonds. *Zero-coupon bonds* (or zero bonds, zeroes or discount bonds) pay only the principal at maturity. No other payments are made during the life of the bond. *Coupon bonds* however promise

[32] See Musiela/Rutkowski (1997), p. 265.
[33] See Shiller (1990), p. 633.
[34] See Martellini/Priaulet/Priaulet (2003), p. 3.
[35] See Fabozzi (2000), p. 3.
[36] See Focardi/Fabozzi (2004), p. 51.
[37] See Fabozzi (2000), p. 4.
[38] See Focardi/Fabozzi (2004), p. 52.

a stream of payments. Like a zero-coupon bond, they pay the principal at the maturity date, but the bondholder also receives interest payments at regular intervals. These intermittent payments are called *coupons*. The amount of the coupon payments is determined by multiplying the coupon rate by the principal.[39] Interest payments are usually made annually or semiannually.

Bond prices are quoted in two different forms.[40] The *dirty price* is the actual amount in return for the right to the full amount of each future coupon payment and the redemption proceeds.

The *clean price* is an artificial price which is, however, the most-quoted price in the marketplace.[41] It is equal to the dirty price minus accrued interest. The *accrued interest* is equal to the amount of the next coupon payment multiplied by the proportion of the current inter-coupon period so far elapsed, i.e. the buyer of the bond "compensates" the seller and pays him that part of the current coupon he already "earned". The popularity of the clean price relies on the fact that it does not fall as a result of a coupon payment.[42]

For further information regarding bond market conventions (day count conventions etc.) see Brown (1994), Stigum/Robinson (1996) and Krgin (2002).

2.2 Interest Rates

The *spot interest rate* $R(t,T)$ designates the rate of interest per period (usually a year) charged on a loan that begins at time t and is paid back at time T, with $t \leq T$.[43] Spot interest rates can be derived from the prices of zero-coupon bonds. By convention we set the zero-coupon bond's principal to 1 unit of account. Let $P(t,T)$ be the price[44] of a zero-coupon bond at time t with maturity T, then there is the following (defining) relationship between zero-coupon bond prices and interest rates[45]

$$P(t,T) = \exp\left(-(T-t)R(t,T)\right) \tag{2.1}$$

We can solve for $R(t,T)$ and obtain

$$R(t,T) = -\frac{1}{T-t}\ln(P(t,T)) \tag{2.2}$$

---

39 With an adjustment for the payment frequency.

40 See Cairns (2004), p. 2.

41 See Cairns (2004), p. 3.

42 See Tuckman (2002), p. 55.

43 The spot interest rate is normally denoted by $r(t,T)$ but we reserve r for the short rate of interest.

44 For zero-coupon bonds the clean and dirty prices are equal since there are no intermediate payments and hence no accrued interest.

45 We assume continuous compounding.

In later chapters we concentrate on modeling one specific spot interest rate: the *short rate*. The short rate is the spot interest rate with instantaneous maturity, i.e.[46]

$$r(t) \equiv \lim_{T \to t} R(t,T) \tag{2.3}$$

The short rate is – of course – unobservable in the market. Therefore a suitable proxy for the short rate has to be found in any empirical work.

The *forward interest rate* ${}_tR(T,\tau)$ is the interest rate set at time t for a loan that begins at time T and is repaid at time τ, with $t \leq T \leq \tau$.[47] It is defined by the following relationship[48]

$$\begin{aligned} \exp(R(t,\tau)(\tau - t)) &= \exp(R(t,T)(T-t)) \exp({}_tR(T,\tau)(\tau - T)) & (2.4)\\ &= \exp(R(t,T)(T-t) +{}_t R(T,\tau)(\tau - T)) & (2.5) \end{aligned}$$

We solve for ${}_tR(T,\tau)$ and obtain

$${}_tR(T,\tau) = \frac{1}{\tau - T} \ln\left(\frac{P(t,T)}{P(t,\tau)}\right) \tag{2.6}$$

which can also be written as[49]

$${}_tR(T,\tau) = R(t,\tau)\frac{(\tau - t)}{(\tau - T)} - R(t,T)\frac{(T-t)}{(\tau - T)} \tag{2.7}$$

Another group of interest rate models to be introduced in a later chapter[50], focuses on a particular set of forward rates namely the *instantaneous forward rates* $f(t,T)$. These are interest rates set at time t for loans that begin at time T and will be repaid an instant later (at time $T+dt$). They are the basic building blocks of the fixed income market since every interest rate can be expressed in terms of instantaneous forward rates. Mathematically they are defined as follows

$$f(t,T) \equiv \lim_{\tau \to T} {}_tR(T,\tau) \tag{2.8}$$

With ${}_tR(T,\tau)$ from (2.6) we obtain the following relationship[51]

$$\begin{aligned} f(t,T) &= \lim_{\tau \to T} \frac{1}{\tau - T} \ln\left(\frac{P(t,T)}{P(t,\tau)}\right)\\ &= -\frac{\partial}{\partial T} \ln P(t,T)\\ &= -\frac{1}{P(t,T)} \frac{\partial P(t,T)}{\partial T} & (2.9) \end{aligned}$$

[46] See Cairns (2004), p. 6.

[47] See Martellini/Priaulet/Priaulet (2003), p. 52.

[48] This relation must hold in an arbitrage-free market. Otherwise one could arbitrage by taking positions in the spot and forward market.

[49] Use equation (2.2).

[50] The Heath/Jarrow/Morton (1992) framework.

[51] See Cairns (2004), p. 5. Note further that $f(t,t) = r(t)$.

Equation (2.9) resembles the modified duration formula for bonds.[52] It can be thought of as the percentage gain in the price of a zero-coupon bond with maturity T if the maturity was to be extended by a small time period.

Zero-coupon bond prices at time t for different maturities T can be calculated from the instantaneous forward rate curve at time t[53]

$$P(t,T) = \exp\left(-\int_t^T f(t,u)\,du\right) \tag{2.10}$$

The *yield to maturity* y_T of a bond maturing at time T is the internal rate of return of that bond, i.e. the interest rate that when used for discounting the future payments results in a net present value of zero. Or in other words, discounting all future payments with the yield to maturity gives the dirty price P of this coupon bond with maturity T, coupon rate c and face value F.

$$P = \sum_{\tau=1}^{T} \frac{cF}{(1+y_T)^{\tau}} + \frac{F}{(1+y_T)^{T}} \tag{2.11}$$

It should be simply regarded as another way of quoting the price of that particular bond. The yield to maturity of a zero-coupon bond maturing at time T equals the spot interest rate for that maturity. For coupon bonds, the yield to maturity is not the actual holding period return of the bond, because this holding period return depends on the reinvestment rates of the coupons. Only if all coupons could be reinvested at the yield to maturity, the holding period return would equal the yield to maturity. The yield to maturity is bond specific and does not give detailed information about the interest rates for specific dates, since it is only a complex average of the respective spot interest rates. The collection of the yield to maturity of all quoted bonds – the yield curve – gives only a crude indication of the interest rates on the market since yields are bond and maturity specific. A better construct (at least for academic purposes) is the term structure of interest rates.

2.3 Term Structure of Interest Rates

The *term structure of interest rates* at a given time is the functional relationship between spot interest rates and term to maturity.[54] It can also be described by forward rates or discount factors, since they contain the same information.[55]

For the term structure to be observable in the market, zero-coupon bonds of different maturities must trade. In reality, only zero-coupon bonds of very few maturities (typically smaller than a year) trade.[56]

---

[52] Modified duration is defined as $\mathrm{MD} = -\frac{1}{P(r)}\frac{\partial P(r)}{\partial r}$.

[53] Solve equation (2.9) for $P(t,T)$.

[54] See Wilhelm (1995), p. 2052.

[55] See Martellini/Priaulet/Priaulet (2003), p. 63.

But an abundance of traded coupon bonds exist in the fixed income markets. Coupon bonds can be thought of as portfolios of zero-coupon bonds. Hence, only specific portfolios of zero-coupon bonds trade.

The problem is then to extract individual zero-coupon prices from the prices of particular portfolios of zero-coupon bonds. From these (theoretical) zero-coupon bond prices the corresponding spot rates can then be calculated.

When the term structure of interest rates is known, one can hence value all assets with known future cash flows, e.g. government bonds.

2.4 Estimating the Term Structure of Interest Rates

In this section we give only a brief overview of term structure estimation techniques since there exists a rich literature on the subject and a full introduction would be outside the scope of this thesis.

Under some very restrictive assumptions on the nature of the traded coupon bonds, the term structure of interest rates can be calculated by the *bootstrap method*.[57] This method uses an iterative process. One starts with a one-period zero-coupon bond[58] and calculates the spot interest rate $R(0,1)$. This interest rate is then used to calculate the spot interest rate $R(0,2)$ given the price of a two-year coupon bond. This method works only if the coupon bonds have the same coupon payment dates.[59] Furthermore, it produces not a smooth term structure but only a collection of interest rates for specific maturities. It is common practice to interpolate the interest rates between two points linearly.[60] These restrictive assumptions are usually not met in the real world, but nevertheless the bootstrap method seems to be quite popular among practitioners.[61]

A further approach is to assume a specific (parameterized) functional form for the term structure (or the discount function).[62] The objective is then to choose the parameters in such a way as to fit the model bond prices as closely as possible to observed bond prices. Two frequently applied parametrization techniques are the *cubic spline method* and the *Nelson-Siegel* approach.[63]

---

56 Recently some governments introduced stripping, i.e. the creation of zero-coupon bonds out of coupon bonds. But the coupon-only and principal-only strips are not as liquid as the coupon bonds and so their prices should not be used to construct the term structure of interest rates, because of the liquidity effect.

57 See Caks (1977), p. 104.

58 Every coupon bond becomes a zero-coupon bond eventually.

59 See Munk (2004b), p. 22.

60 See Hull (2005), p. 84.

61 Bootstrapping is also used for the derivation of the swap curve. It is common to use money market rates, futures and swap rates for its construction.

62 With some additional assumptions one therefore obtains a smooth term structure.

63 See Munk (2004b), p. 21.

The cubic spline method was first introduced by McCulloch (1971). This method assumes that the maturity axis is divided into subintervals and that separate functions of the same type (so-called splines) are used to describe the discount function in the different subintervals.[64] The problem with this method is that the derived forward rate curve will usually be quite rugged and that the curve is quite sensitive to the bond prices and the location of the knot points.[65]

Nelson/Siegel (1987) propose a parsimonious model of the term structure that has become quite popular particularly among central banks. The model was later extended by Svensson (1994). Their parametrization of the instantaneous forward rate curve allows for quite flexible shapes of the term structure of interest rates although it relies on only 4 parameters. Furthermore the forward rate curve is smooth.[66]

For further information about term structure estimation techniques see Chapter 4 in Martellini/Priaulet/Priaulet (2003), Chapters 10 and 11 in Choudhry (2004) and the book by Anderson et al. (1996).

2.5 Classical Theories of the Term Structure of Interest Rates

After the estimation of the term structure the natural question arises of what determines its shape.[67] There exist several yield curve theories that try to answer this question. We only give a short overview here since we will adopt a different approach in this thesis. It is intended to highlight the differences between the classical theories and modern arbitrage-free theories of the term-structure of interest rates.

The *expectations theory* was developed by Hicks (1939) and Lutz (1940). It has a number of variants all relying on the idea that current interest rates are linked to expected future rates. For a critical examination see Cox/Ingersoll/Ross (1981).

The *liquidity preference theory* introduced by Hicks (1939) finds that the expectations theory ignores the investor's risk aversion and argues that expected returns on bonds with longer maturities should be higher than for shorter bonds to compensate for the higher price fluctuation of longer bonds.[68] According to this hypothesis, the term structure of interest rates must be normal, i.e. spot rates rise with longer maturity.

In contrast, the *market segmentation theory* by Culbertson (1957) claims that investors want to invest in an appropriate set of bonds and maturity

---

[64] See Munk (2004b), p. 25.

[65] See Munk (2004b), p. 29.

[66] This is due to the fact, that the functional form of the forward rate curve is an input (assumption) of the model unlike in the cubic spline model.

[67] Traditionally one distinguishes normal, flat and inverse term structure shapes.

[68] See Munk (2004b), p. 117.

segments that are suitable for their purpose.[69] Since different investors act in different ways there is no reason to assume that the movement of interest rates in different maturity segments is interrelated.

A more realistic version of this hypothesis is the *preferred habitat theory* introduced by Modigliani/Sutch (1966). They argue that although each investor might prefer to invest in a particular set of bonds, he should be willing to invest in different bonds if he is sufficiently compensated for doing so.[70] The different segments of the bond market are therefore not completely independent of each other.

2.6 Arbitrage-Free Term Structure Theories

In this thesis, the classical theories are not considered because of their shortcomings regarding the pricing of bonds. We consider arbitrage free pricing theories which pull together the classical theories in a mathematical precise way.[71] These theories will be presented in Chapter 3.

2.7 Empirical Properties of the Term Structure of Interest Rates

Before we turn to the topic of (arbitrage-free) interest rate modeling, we need to take a closer look at some empirical properties of the term structure. A realistic model should at least possess some of these properties. Rebonato (1998) describes some features of desirable interest rate models.[72] Interest rates should not become negative and display mean reversion.[73] Mean reversion refers to a level-dependent drift of a stochastic process, i.e. the drift is positive (negative) when the current value of the stochastic process is below (above) a certain level.[74] The correlation between interest rates of different maturities generated by the model should be positive but imperfect.[75] The term structure of volatility is normally not flat but humped.[76] In a humped volatility structure, the volatility first rises and then declines.

As usual, model choice is a trade-off between realism and (analytical or numerical) tractability. In this thesis, we focus more on tractability. In Chapter 3, we will present the general framework for term structure modeling and derive two analytically tractable models as special cases.

---

[69] See Cairns (2004), p. 12.
[70] See Munk (2004b), p. 118.
[71] See Cairns (2004), p. 13.
[72] See Rebonato (1998), pp. 233–234.
[73] See Rebonato (1998), p. 233.
[74] The stochastic process for the short rate in the Vasicek term structure model to be introduced in Chapter 3 possesses this feature.
[75] See Golub/Tilman (2000), p. 89.
[76] See Golub/Tilman (2000), p. 89.

3

Term Structure Modeling in Continuous Time

3.1 Introduction

The main inputs of portfolio selection models are the expected values and covariances of the assets under consideration. In equity portfolio selection, the expected values and covariances are oftentimes estimated by analyzing the historical time series of the stocks. Because of bond characteristics and properties of bond portfolio selection models that will be discussed in greater detail in Chapter 4, such an approach is generally ruled out for fixed income instruments. In order to determine the bond portfolio selection parameters consistently, a theoretical model for the evolution of bond prices over time is needed.[77]

We introduced the term structure of interest rates in Chapter 2. Given the term structure at time t, the time t value of all bonds with known future payoffs can be calculated.[78] Furthermore, future term structures determine future bond prices. Bonds can therefore be regarded as term structure (or interest rate) derivatives since their value depends on interest rates. Hence, a theory about the dynamics of the term structure yields at the same time a model for the movements of bonds over time.[79]

Dynamic term structure modeling – i.e. modeling the evolution of the term structure of interest rates over time in an arbitrage-free fashion – is one of the most heavily researched areas in financial economics.[80]

An early attempt to model the term structure dynamics was the duration model.[81] It assumes that the term structure of interest rates is flat and moves only in a parallel fashion. On first thought this seems to be a reasonable assumption, but it turns out that this is problematic for two reasons: (i)

[77] See Wilhelm (1992), p. 213.
[78] See Wilhelm (1992), p. 209.
[79] See Wilhelm (1992), p. 209.
[80] Choudhry (2004), p. 178.
[81] The model dates back to the seminal analysis of Macaulay (1938).

such term structure movements are empirically improbable and (ii) if such movements would occur, it would present an arbitrage opportunity.[82] The second argument is of course much more serious from a theoretical point of view.

Using the duration model for bond portfolio management purposes is therefore unreasonable. Due to the presence of arbitrage opportunities, there would not exist an optimal solution to the portfolio selection problem, in the sense that every possible portfolio would be dominated. Consequently – at least for bond portfolio selection purposes – there is a need for better interest rate models.[83]

It is generally accepted that term structure modeling theory started with the seminal paper of Vasicek (1977). Since then a vast number of interest rate models have been proposed by academics and practitioners alike.

Section 3.2 gives an overview of possible term structure modeling approaches. In Section 3.3 we introduce the Heath/Jarrow/Morton (1992) framework and derive a general zero-coupon bond pricing equation using stochastic discount factors. Sections 3.4 and 3.5 derive two special cases of the general Heath/Jarrow/Morton (1992) framework, that will be used in later chapters on bond portfolio optimization.

3.2 Interest Rate Modeling Approaches

There are several approaches to modeling the term structure of interest rates. A possible classification distinguishes between whole yield curve models and short rate models.[84]

Whole Yield Curve Models

The most general approach is to specify the stochastic processes followed by instantaneous forward rates $f(t,T)$.[85] For a fixed maturity T, we can write this process as

$$df(t,T) = m(t,T)dt + s_1(t,T)dz_1$$

where $m(t,T)$ is the drift, $s_1(t,T)$ is the instantaneous standard deviation and dz_1 is a Brownian motion. This approach has been proposed in a groundbreaking paper by Heath/Jarrow/Morton (1992) (hereafter HJM). HJM model the whole term structure of instantaneous forward rates, i.e. an infinite number of interest rates. They show that the drift of the forward rates must follow

[82] See Cairns (2004), p. 18.

[83] See Baz/Chacko (2004), p. 108.

[84] For ease of exposition we assume in this section that a single Brownian motion drives the whole term structure.

[85] Instantaneous forward rates have been introduced in Equation (2.9).

from the specification of the volatilities and the market prices of interest rate risk.[86]

Hull/White (1996) show that any model of zero-coupon bond prices can be converted into an equivalent model of instantaneous forward rates and vice versa.[87] The dynamics of the zero-coupon bond prices can be written as[88]

$$\frac{dP(t,T)}{P(t,T)} = \mu(t,T)dt + \sigma_1(t,T)dz_1$$

where $P(t,T)$ is the time t price of a zero-coupon bond with maturity T, $\mu(t,T)$ is the drift and $\sigma_1(t,T)$ is the bond price volatility. It has been shown that in an arbitrage-free setting, the drift of the bond prices is a particular function of the short rate r, the volatilities σ_1 and the market prices of interest rate risk. The volatility of the bond price must be a function of maturity since it must decline as maturity approaches. Since the price of the bond at maturity is known with certainty, we must have that the volatility at maturity is zero, i.e. $\sigma_1(T,T) = 0$.[89]

The advantage of specifying a model in terms of the processes followed by either bond prices or forward rates is that the model is automatically consistent with the initial term structure, since this initial term structure determines the initial values of the variables being modeled.[90] The disadvantage is that the model is usually non-Markov[91] and therefore slow computationally.[92] As has been shown, a forward rate process is structurally simpler than a bond price process in that it usually does not depend on the short rate r.[93]

Models of the Short Rate

The most widely used approach is to model the evolution of one instantaneous forward rate only, namely the short rate $r(t)$.[94] The economic intuition behind short rate models arises from the observation that bond prices tend to move together, i.e. changes in bond prices are highly correlated.[95] Short rate

[86] See Chapter 3.3.4 for a derivation of this drift condition.
[87] Hull/White (1996), pp. 261–262.
[88] See Hull/White (1996), p. 228.
[89] See Hull/White (1996), p. 229.
[90] See Hull/White (1996), p. 229.
[91] A stochastic process has the Markov property if the conditional distribution of its future values depends only on the current value and not on the past. See Cvitanic/Zapatero (2004), p. 65.
[92] See Hull/White (1996), p. 229.
[93] Hull/White (1996), p. 229.
[94] It has been common to assume that only one factor drives the term structure. According to Martellini/Priaulet/Priaulet (2003), p. 388, there is a general consensus among researchers that this factor should be the short rate.
[95] See Table 3.1 on page 70 of Martellini/Priaulet/Priaulet (2003) for empirical results of interest-rate co-movements.

changes are thus used as a proxy for changes in the level of the whole term structure.[96] Empirical studies – for example Litterman/Scheinkman (1991) – confirmed that level changes account for a large part of the dynamics of the term structure.[97] Short rate models propose a diffusion process for the evolution of the short rate[98]

$$dr(t) = \alpha(r,t)dt + \sigma_r(r,t)dz_1 \tag{3.1}$$

where $\alpha(r,t)$ is the drift of the short rate and $\sigma_r(r,t)$ is its volatility. The dynamics of other interest rates is derived endogenously. Since α and σ_r only depend on the value of the short rate, the model is always Markov.[99] This model is automatically consistent with the current value of the short rate but not necessarily with other interest rates.[100] In other words, because only the dynamics of the short rate is specified, the only exogenously given asset is the money market account[101] and zero-coupon bond prices are considered derivatives on the short rate.[102]

Whole Yield Curve Models versus Short Rate Models

Whole yield curve models can be differentiated from the short rate models with regard to various criteria[103]

- *Traded assets.* The only traded asset in the short rate models is the money market account, zero-coupon bonds are regarded as derivatives on the short rate. The HJM framework assumes that zero-coupon bonds of all maturities trade (including one with instantaneous maturity, i.e. the money market account).
- *Completeness.* A market is said to be complete if any contingent claim can be replicated with existing securities.[104] Short rate models are not complete. , because there is (at least) one source of risk and just one traded asset (the money market account). HJM models are complete, because there are d risk sources and an infinite number of traded assets.
- *Freedom of arbitrage.* Short rate models are arbitrage-free per construction, because the only traded asset (and therefore the only possible portfolio) is the money market account. Furthermore since holding of cash is prohibited, negative interest rates don't lead to arbitrage opportunities.

---

96 See Martellini/Priaulet/Priaulet (2003), p. 388.
97 See Martellini/Priaulet/Priaulet (2003), p. 388.
98 Svoboda (2004), p. vii.
99 See Hull/White (1996), p. 229.
100 See Hull/White (1996), p. 229.
101 The money market account or cash account yields the current short rate, see Chapter 3.3.2.
102 See Björk (1998), p. 242.
103 See Branger/Schlag (2004), p. 160.
104 See Cvitanic/Zapatero (2004), p. 88.

The HJM framework is only free of arbitrage opportunities if the drift condition[105] is met.

- *Matching of the initial term structure.* Time-homogeneous short rate models[106] cannot generally fit the current term structure because a finite number of parameters cannot be chosen in such a way as to fit an infinite number of prices (or interest rates). Time-inhomogeneous short rate models solve this problem by making some (or all) parameters time-dependent. The HJM model fits the current yield by construction because it is taken as an input.
- *Market prices of risk.* In a short rate model, the drift and volatility of the short rate and market price of risk must be specified. It is particularly difficult to estimate the drift of the short rate. In a HJM model the drift of the interest rates follow from arbitrage considerations and therefore only the current term structure, the volatilities of the forward rates and the market prices of risk must be known.

In the following sections we present the general HJM framework and derive two well-known term structure models as special cases, the Vasicek (1977) model and the Hull/White (1994) two-factor model.

3.3 Heath/Jarrow/Morton (1992)

3.3.1 Introduction

The general framework for arbitrage-free interest rate modeling in continuous time was proposed in a pioneering paper by HJM. The HJM framework is a multi-factor model, i.e. the term structure of interest is subject to multiple shocks. It can be shown that all previously developed[107] interest rate models are special cases of this framework.[108]

HJM start with the observed term structure of interest rates in its instantaneous forward rate form.[109] For a fixed but arbitrary maturity $T \in [t, \tau]$, $f(t,T)$ satisfies[110]

$$df(t,T) = m(t,T)dt + \sum_{i=1}^{d} s_i(t,T)dz_i(t) \tag{3.2}$$

where $z_1, \dots, z_d$ are d independent standard Brownian motions, $m(t,T)$ is the drift and $s_i(t,T)$ is the i-th volatility of $f(t,T)$. In its most general form, both

[105] To be derived in Chapter 3.3.4.

[106] If both the drift and the volatility are independent of time, the diffusion is said to be time-homogeneous, otherwise it is said to be time-inhomogeneous.

[107] This includes all short rate models.

[108] Svoboda (2004), p. 124.

[109] See Branger/Schlag (2004), p. 126.

[110] Heath/Jarrow/Morton (1992), p. 80.

the drift and the volatility may depend on the entire forward rate curve at time t.[111]

Consequently, the whole term structure of interest rates – an infinite number of stochastic processes – is modeled. But since the model has only finite Brownian motions as risk sources, there must be some kind of arbitrage-free condition for the forward rate drift (the HJM drift condition).[112] This drift condition ensures that the traded zero-coupon bonds form an arbitrage-free market.[113]

This section is organized as follows. Section 3.3.2 derives the dynamics of traded zero-coupon bond prices that must follow from the specification of the instantaneous forward rates in Equation (3.2). In Section 3.3.3 we show how the prices of interest rate derivatives can be obtained by employing the stochastic discount factor approach. In Section 3.3.4 we derive the HJM drift condition and Section 3.3.5 derives the short rate dynamics that provides a formal link to the special cases (short rate models) introduced later in the chapter.

3.3.2 Dynamics of Traded Securities

In the HJM framework, it is assumed that a money market account $B(t)$ and zero-coupon bonds $P(t,T)$ of different maturities T trade.[114]

The dynamics of the money market account

$$B(t) = B(0) \exp\left(\int_0^t f(u,u)du\right), \quad B(0) = 1$$

can be obtained by a straightforward application of Itô's lemma

$$\frac{dB(t)}{B(t)} = f(t,t)dt \tag{3.3}$$

With Equation (2.10) we can derive zero-coupon bond prices from the instantaneous forward rate curve

$$P(t,T) = \exp\left(-\int_t^T f(t,u)du\right) \tag{3.4}$$

Applying the stochastic Fubini theorem[115] and Itô's lemma, we obtain the dynamics of zero-coupon bonds[116]

[111] See Heath/Jarrow/Morton (1992), p. 80. For notational convenience this dependence is not indicated in Equation (3.2).

[112] See Branger/Schlag (2004), p. 126.

[113] We derive the drift condition in Chapter 3.3.4.

[114] See Heath/Jarrow/Morton (1992), p. 79.

[115] See Heath/Jarrow/Morton (1992), p. 99.

[116] For a derivation see Appendix A.1.

$$\frac{dP(t,T)}{P(t,T)} = \mu(t,T)dt - \sum_{i=1}^{d} \sigma_i(t,T)dz_i(t) \tag{3.5}$$

where

$$\mu(t,T) = f(t,t) - \int_t^T m(t,u)du + \frac{1}{2}\sum_{i=1}^{d}\left(\int_t^T s_i(t,u)du\right)^2 \tag{3.6}$$

$$\sigma_i(t,T) = \int_t^T s_i(t,u)du \tag{3.7}$$

3.3.3 Arbitrage-Free Pricing

The prices of interest rate derivative securities in an arbitrage-free market can be obtained by multiple techniques. Originally, the derivation of the interest rate models followed either the PDE approach or the martingale approach.[117] In this thesis, we use the stochastic discount factor methodology for pricing interest-rate derivative securities.[118]

The fundamental theorem of asset pricing states that the time t price $P(t)$ of a financial claim can be obtained in an arbitrage-free market by taking the expectation with respect to the real-world probabilities over the product of stochastic discount factor ζ and payoff $P(T)$[119]

$$P(t) = E_t\left[\frac{\zeta_T}{\zeta_t}P(T)\right] \tag{3.8}$$

We now assume the following dynamics for the stochastic discount factor[120]

$$d\zeta(t) = a(t)\zeta(t)dt + \zeta(t)\sum_{i=1}^{d} b_i(t)dz_i(t)$$

where $a(t)$ is the drift and $b_i(t)$ is the volatility of the stochastic discount factor with regard to the i-th Brownian motion.

For the fixed income market to be free of arbitrage, the drift and volatilities of the stochastic discount factor cannot be chosen arbitrarily. Jin/Glasserman

[117] For a brief exposition of the two approaches see Cairns (2004), p. 55 and Cairns (2004), p. 60.

[118] For a book-length treatment of asset pricing that contains a chapter on the stochatic discount factor methodology see Cochrane (2005). Wilhelm (2005) derives Gaussian interest rate models from the specification of the stochastic discount factor. The Ho-Lee interest rate model was derived by stochastic discounting already in Wilhelm (1999). For stochastic discounting in a discrete time setting see Wilhelm (1996).

[119] If the market is free of arbitrage and complete, there exists a unique stochastic discount factor, see Harrison/Kreps (1979) and Harrison/Pliska (1981).

[120] Based on the one-factor formulation in Baz/Chacko (2004), p. 51.

(2001) show that the drift and the volatilities of the stochastic discount factor must be related to the short rate $f(t,t)$ and the market prices of interest rate risk $\lambda_i(t)$[121] in the following way[122]

$$a(t) = -f(t,t) \tag{3.9}$$
$$b_i(t) = \lambda_i(t) \; \forall \, i = 1, \ldots, d \tag{3.10}$$

The drift of the stochastic discount factor must be equal to the negative short rate and the volatilities must be equal to the market prices of interest rate risk. The dynamics of the stochastic discount factor can then be given as

$$d\zeta(t) = -f(t,t)\zeta(t)dt + \zeta(t)\sum_{i=1}^{d} \lambda_i(t)dz_i(t) \tag{3.11}$$

Equation (3.11) can be solved explicitly.[123] We obtain

$$\zeta(T) = \zeta(t)\exp\left(\int_t^T -f(u,u)du + \sum_{i=1}^{d}\int_t^T \lambda_i(u)dz_i(u) - \sum_{i=1}^{d}\frac{1}{2}\int_t^T \lambda_i(u)^2 du\right)$$

and the arbitrage-free prices of zero-coupon bonds can be derived with the following pricing equation using (3.8) and $P(T,T) = 1$

$$P(t,T) = E_t\left[\exp\left(-\int_t^T f(u,u)du + \sum_{i=1}^{d}\int_t^T \lambda_i(u)dz_i(u) - \sum_{i=1}^{d}\frac{1}{2}\int_t^T \lambda_i(u)^2 du\right)\right] \tag{3.12}$$

The valuation of zero-coupon bonds hence requires a specification of the market prices of interest rate risk λ_i and of the short rate evolution. Before we derive an expression for the short rate, we derive the HJM drift condition.

3.3.4 Excursus: The HJM Drift Condition

The groundwork laid in the last sections allows us to derive the classical HJM drift condition. First, we calculate the deflated price process $Y(t,T)$ with

$$Y(t,T) = \zeta(t)P(t,T)$$

[121] The market price of interest rate risk gives the expected return over the riskless rate per unit of volatility. It must be equal for all traded assets, see Branger/Schlag (2004), p. 115.

[122] See Jin/Glasserman (2001), p. 195.

[123] For a derivation see Appendix A.2.

The deflated price process is a martingale.[124] An application of Itô's lemma with $dP(t,T)$ from (3.5) and $d\zeta$ from (3.11) yields the dynamics of Y[125]

$$\frac{dY(t,T)}{Y(t,T)} = \left(-f(t,t) + \mu(t,T) - \sum_{i=1}^{d} \lambda_i(t)\sigma_i(t,T)\right) dt + \sum_{i=1}^{d} (\lambda_i(t) - \sigma_i(t,T)) dz_i(t)$$

Since $Y(t,T)$ is a martingale, the drift of $Y(t,T)$ must be equal to zero. We obtain

$$\mu(t,T) = f(t,t) + \sum_{i=1}^{d} \sigma_i(t,T)\lambda_i(t) \tag{3.13}$$

We insert $\mu(t,T)$ and $\sigma_i(t,T)$ from (3.6) and (3.7) into (3.13) and obtain

$$\int_t^T m(t,u)du = \frac{1}{2}\sum_{i=1}^{d} \left(\int_t^T s_i(t,u)du\right)^2 - \sum_{i=1}^{d} \lambda_i \left(\int_t^T s_i(t,u)du\right)$$

Differentiating this equation with respect to T yields the so-called HJM drift condition[126]

$$m(t,T) = \sum_{i=1}^{d} s_i(t,T) \left(\left(\int_t^T s_i(t,u)du\right) - \lambda_i(t)\right) \tag{3.14}$$

The drift of the forward rates cannot be chosen independently but it results from the specification of the volatility structure of the forward rates and the market prices of risk.[127]

3.3.5 The Short Rate of Interest

The short rate is defined as the instantaneous forward rate with instantaneous maturity, i.e. it can be obtained as the limit $T \to t$ of the instantaneous limit forward rate function $f(t,T)$. The integrated dynamics of $f(t,T)$ can be derived with (3.2) and (3.14). It follows that

$$\begin{aligned} f(t,T) =& f(0,T) + \sum_{i=1}^{d} \int_0^t s_i(u,T) \left(\left(\int_u^T s_i(u,s)ds\right) - \lambda_i(u)\right) du \\ &+ \sum_{i=1}^{d} \int_0^t s_i(u,T)dz_i(u) \end{aligned}$$

[124] This follows from the definition of the stochastic discount factor, see Duffie (1996), p. 103. A stochastic process X is a martingale if $E_t(X_s) = X_t$, see Duffie (1996), p. 22.

[125] For a derivation see Appendix A.3.

[126] See Heath/Jarrow/Morton (1992), p. 84.

[127] See Branger/Schlag (2004), p. 129.

We obtain the short rate by letting $T \to t$

$$
\begin{aligned}
f(t,t) =& f(0,t) + \sum_{i=1}^{d} \int_0^t s_i(u,t) \left(\left(\int_u^t s_i(u,s)ds \right) - \lambda_i(u) \right) du \\
&+ \sum_{i=1}^{d} \int_0^t s_i(u,t) dz_i(u)
\end{aligned} \tag{3.15}
$$

The time t short rate hence depends on the initial instantaneous forward rate $f(0,t)$, the market prices of interest rate risk and the forward rate volatilities.

3.3.6 Special Cases

Particular interest rate models can be derived by applying the general pricing equation (3.12) with $f(t,t)$ from (3.15) and specifying

- volatility function(s) of the instantaneous forward rates $s_i(t,T)$
- functional form of the initial instantaneous forward rate curve $f(0,t)$
- the market price(s) of risk $\lambda_i(t)$

In the next two sections, we introduce two special cases of the general HJM interest rate framework

- One-factor Vasicek (1977) model
- Two-factor Hull/White (1994) model

We chose these models because they are quite realistic term structure models and – of major importance for our analysis – analytically tractable. The Vasicek model can be obtained from the general HJM framework by setting[128]

Table 3.1. Vasicek (1977) model: Special case of HJM.

Item	Vasicek (1977) model
$s(t,T)$	$\sigma_r \exp(-\kappa(T-t))$
$\lambda(t)$	λ
$f(0,T)$	$\theta + \exp(-\kappa T)(f(0,0) - \theta) + \lambda \frac{\sigma_r}{\kappa}(1 - \exp(-\kappa T)) - \frac{\sigma_r^2}{2\kappa^2}(1 - \exp(-\kappa T))^2$

In a similar fashion, the HW2 model can be derived by setting[129]

[128] A derivation of the model follows in the next part of the chapter.

[129] With $g(0,T)$ defined in Equation (A.6) in Appendix A.4. A derivation of the model follows in the next part of the chapter.

Table 3.2. Hull/White (1994) model: Special case of HJM.

Item	Hull/White (1994) model
$s_1(t,T)$	$\frac{e^{(t-T)\kappa_r}(\kappa_r-\kappa_\varepsilon)\sigma_r-\left(e^{(t-T)\kappa_r}-e^{(t-T)\kappa_\varepsilon}\right)\varrho\sigma_\varepsilon}{\kappa_r-\kappa_\varepsilon}$
$s_2(t,T)$	$\frac{\left(-e^{(t-T)\kappa_r}+e^{(t-T)\kappa_\varepsilon}\right)\sqrt{1-\varrho^2}\sigma_\varepsilon}{\kappa_r-\kappa_\varepsilon}$
$\lambda_1(t)$	λ_1
$\lambda_2(t)$	λ_2
$f(0,T)$	$g(0,T)+r(0)e^{-\kappa_r T}+\varepsilon(0)\frac{e^{-\kappa_\varepsilon T}-e^{-\kappa_r T}}{\kappa_r-\kappa_\varepsilon}$

In chapters 3.4 and 3.5 we will derive and analyze these special cases of the general HJM interest rate modeling framework.

3.4 Vasicek (1977)

3.4.1 Introduction

In 1977 Vasicek proposed what is known to be the first arbitrage-free dynamic interest rate model.[130] He assumed that the term structure of interest rates is completely determined by the current value of only one random variable – the short rate of interest. The short rate follows an Ornstein-Uhlenbeck process. This is a stationary Markov process with normally distributed increments.[131] The behavior of the short rate can be described by the following SDE[132]

$$dr(t) = \kappa(\theta - r(t))dt + \sigma_r dz_1(t) \tag{3.16}$$

where $\kappa > 0$, θ and σ_r are constant.[133] θ designates the mean reversion level, κ is the reversion speed and σ_r is the volatility of the short rate. An Ornstein-Uhlenbeck process exhibits mean reversion. The drift is positive when $r(t) < \theta$ and negative when $r(t) > \theta$. The process is therefore pulled towards θ. The magnitude of the pull depends on the reversion speed κ.

3.4.2 Derivation of Zero-Coupon Bond Prices

In Chapter 3.3.3 we obtained a general pricing equation for zero-coupon bonds.[134] We now want to specify the initial forward rate curve, volatility structure and the market prices of interest rate risk in order to obtain a formula for zero-coupon bond prices in the Vasicek (1977) model.

---

[130] Vasicek (1977).
[131] Vasicek (1977), p. 185.
[132] Vasicek (1977), p. 185.
[133] Vasicek (1977), p. 185.
[134] See Equation (3.12).

The specifications for the Vasicek (1977) model in the HJM framework of Chapter 3.3 are as follows. The model is a one-factor interest rate model, i.e. $d = 1$. The forward rate volatilities are assumed to be of the following form

$$s_1(t,T) = s(t,T) = \sigma_r \exp(-\kappa(T-t)) \tag{3.17}$$

and the market price of interest rate risk is a constant, i.e.

$$\lambda_1(t) = \lambda(t) = \lambda \tag{3.18}$$

The initial instantaneous forward rate curve is given by

$$\begin{aligned} f(0,T) =& \theta + \exp(-\kappa T))(f(0,0) - \theta) \\ &+ \lambda\frac{\sigma_r}{\kappa}(1-\exp(-\kappa T)) - \frac{\sigma_r^2}{2\kappa^2}(1-\exp(-\kappa T))^2 \end{aligned} \tag{3.19}$$

With these specifications, Equation (3.15) determines the short rate in the Vasicek model. We obtain[135]

$$\begin{aligned} r(T) &= f(T,T) \\ &= f(t,T) + \int_t^T s(u,T)\left(\int_u^T s(u,s)ds - \lambda\right)du + \int_t^T s(u,T)dz(u) \\ &= r(t)e^{-\kappa(T-t)} + \theta(1-e^{-\kappa(T-t)}) + \sigma_r \int_t^T e^{-\kappa(T-u)}dz(u) \end{aligned} \tag{3.20}$$

Now, we are in a position to compute zero-coupon bond prices, since the general bond pricing equation (3.12) is completely specified. It follows that

$$\begin{aligned} P(t,T) = E_t\Bigg[\exp\Bigg(&-r(t)\frac{1-e^{-\kappa(T-t)}}{\kappa} - \theta\frac{(-1+e^{-\kappa(T-t)}+\kappa(T-t))}{\kappa} \\ &-\frac{1}{2}\lambda^2(T-t) + \frac{\sigma_r}{\kappa}\int_t^T ((e^{-\kappa(T-s)}-1)+\lambda)dz(s)\Bigg)\Bigg] \end{aligned} \tag{3.21}$$

There is only one stochastic expression in Equation (3.21), hence

$$\begin{aligned} P(t,T) = \exp\Bigg(&-r(t)\frac{1-e^{-\kappa(T-t)}}{\kappa} - \theta\frac{(-1+e^{-\kappa(T-t)}+\kappa(T-t))}{\kappa} \\ &-\frac{1}{2}\lambda^2(T-t)\Bigg) E_t\left[\exp\underbrace{\left(\frac{\sigma_r}{\kappa}\int_t^T ((e^{-\kappa(T-s)}-1)+\lambda)dz(s)\right)}_{x}\right] \end{aligned} \tag{3.22}$$

135 This could be obtained as well by solving (3.16).

Since x is a normally distributed random variable, it follows that[136]

$$E[e^x] = \exp\left(E[x] + \frac{1}{2}\mathrm{var}(x)\right)$$
$$\mathrm{var}(e^x) = E[e^x]^2(\exp(\mathrm{var}(x)) - 1)$$

Hence, we obtain

$$\begin{aligned} &E_t\left[\exp\left(\frac{\sigma_r}{\kappa}\int_t^T ((e^{-\kappa(T-s)} - 1) + \lambda)dz(s)\right)\right] \\ &= \exp\left(\frac{1}{2}\mathrm{var}\left[\int_t^T \left(\frac{\sigma_r}{\kappa}((e^{-\kappa(T-s)} - 1) + \lambda)\right)^2 dz(s)\right]\right) \\ &= \exp\left(\frac{1}{2}\int_t^T \left(\frac{\sigma_r}{\kappa}((e^{-\kappa(T-s)} - 1) + \lambda)\right)^2 ds\right) \\ &= \exp\left(-\frac{\left(3 + e^{-2(T-t)\kappa} - 4e^{-(T-t)\kappa}\right)\sigma_r^2}{4\kappa^3} + \frac{\left(1 - e^{-(T-t)\kappa}\right)\lambda\sigma_r}{\kappa^2}\right. \\ &\left. + \frac{(T-t)(\sigma_r - \kappa\lambda)^2}{2\kappa^2}\right) \end{aligned} \tag{3.23}$$

By substituting (3.23) into (3.22), we eventually obtain the zero-coupon bond price formula

$$\begin{aligned} P(t,T) = \exp&\left(-\frac{1}{2}(T-t)\lambda^2 + \frac{\left(1 - e^{-(T-t)\kappa}\right)\sigma_r\lambda}{\kappa^2}\right. \\ &- \frac{\left(3 + e^{-2(T-t)\kappa} - 4e^{-(T-t)\kappa}\right)\sigma_r^2}{4\kappa^3} + \frac{(T-t)(\sigma_r - \kappa\lambda)^2}{2\kappa^2} \\ &\left. - \frac{\theta\left((T-t)\kappa + e^{-(T-t)\kappa} - 1\right)}{\kappa} - \frac{\left(1 - e^{-(T-t)\kappa}\right)}{\kappa}r(t)\right) \end{aligned} \tag{3.24}$$

This can be written in a shorter way by setting[137]

$$B(t,T) = \frac{1}{\kappa}(1 - \exp(-\kappa(T-t))) \tag{3.25}$$

and[138]

$$\begin{aligned} A(t,T) = &R(\infty)\left(\frac{1}{\kappa}(1 - \exp(-\kappa(T-t))) - (T-t)\right) \\ &- \frac{\sigma_r^2}{4\kappa^3}(1 - \exp(-\kappa(T-t)))^2 \end{aligned} \tag{3.26}$$

[136] See Rinne (1997), p. 365.
[137] Svoboda (2004), p. 10.
[138] Svoboda (2004), p. 10.

where[139]

$$R(\infty) = \left(\theta + \lambda\frac{\sigma_r}{\kappa} - \frac{1}{2}\frac{\sigma_r^2}{\kappa^2}\right) \tag{3.27}$$

Hence, the analytic solution for zero-coupon bond prices in the Vasicek term structure model is

$$P(t,T) = \exp(A(t,T) - B(t,T)r(t)) \tag{3.28}$$

3.4.3 Properties

First, we want to derive some properties of the short rate. It can be seen from (3.16), that the drift term is not constant but depends on the current level of the short rate. θ is the long-term mean level of the short rate. Whenever the current short rate is above (below) the long-term mean level, the drift is negative (positive). The parameter κ – the speed of adjustment – determines how "fast" the short rate is pulled towards θ. Because of this behavior of the short rate, all interest rates in the Vasicek model exhibit mean reversion.[140]

The future short rate $r(T)$ is (conditionally) normally distributed with mean[141]

$$E_t[r(T)] = r(t)\underbrace{\exp(-\kappa(T-t))}_{w} + \theta\underbrace{(1-\exp(-\kappa(T-t)))}_{1-w} \tag{3.29}$$

and variance[142]

$$\text{var}_t(r(T)) = \frac{\sigma_r^2}{2\kappa}(1-\exp(-2\kappa(T-t))) \tag{3.30}$$

Expected future short rates are a weighted average of the current value of the short rate $r(t)$ and the long-term mean value θ, the weights being w and $1-w$ respectively.[143] For small values of κ the short rate reverts slowly to the long-term mean and therefore more weight is given to the current value of the short rate. For $\kappa \to 0$, the weight w becomes 1 and hence the expected value equals the current value.[144] For $T \to \infty$ (and any value of κ) the mean approaches θ.[145]

The effect of the parameters σ_r and κ on the variance are easily calculable. An increase (decrease) of σ_r increases (decreases) the variance of future short

139 Vasicek (1977), p. 186.
140 Svoboda (2004), p. 9.
141 Apply the expectations operator to Equation (3.20).
142 This can be derived as well from Equation (3.20). See also Vasicek (1977), p. 185.
143 See Tuckman (2002), p. 239.
144 For large values of κ the short rate reverts fast to the long-term mean and therefore less weight is given to the current value of the short rate. For $\kappa \to \infty$, the weight w becomes zero and hence the expected value equals the long-term mean level θ.
145 Munk (2004b), p. 157.

rates and an increase (decrease) in κ decreases (increases) it. This is quite obvious, since an increase of κ makes the short rate revert faster to the long-term mean level θ and so the variance decreases.

A further inspection of (3.29) reveals the effect of the mean reversion feature on short rate expectations. Depending on the relation between the current short rate and the long-term mean level, the following cases can be distinguished

- $E_t[r(T)] = r(t)$ if $r(t) = \theta$. Whenever the current short rate is equal to its long term mean θ, the expectation is, that it doesn't change,[146] since over the long-term it must be pulled towards θ and this expected path is disturbed only by uncorrelated zero-mean shocks.
- $E_t[r(T)] < r(t)$ if $r(t) > \theta$. Whenever the current short rate is above θ, it is expected to decline.
- $E_t[r(T)] > r(t)$ if $r(t) < \theta$. Whenever the current short rate is below θ, it is expected to rise.

The Vasicek model is Gaussian. So, like every Gaussian model, it assigns positive probabilities to negative values of the future short rate (and all other future interest rates).[147] Theoretically, this is a undesirable property of a model, since negative interest rates are not possible in an arbitrage free market.[148] The model is nevertheless free of arbitrage since holding cash is not allowed in the model.[149] For "reasonable" parameter values, the possibility of negative future interest rates is fortunately quite small.[150] In an influential empirical study of the term structure Chan et al. (1992) found the following parameters for the Vasicek model: $\kappa = 0.1779$, $\theta = 0.0865$ and $\sigma_r = 0.02$.[151]

For these values, the probability of negative short rates is generally under 5 %[152]

Table 3.3. Vasicek model: Probability of negative future short rates.

T	1	2	3	4	5
$Pr(r(T) < 0)$	4.6 %	4.8 %	4.0 %	3.3 %	2.6 %

[146] The drift term is equal to zero.

[147] See Martellini/Priaulet/Priaulet (2003), p. 390.

[148] People could hold cash and getting a zero return instead.

[149] Munk (2004b), p. 159.

[150] It is $N\left(-\frac{E_t[R(T)]}{\sqrt{\mathrm{var}_t(r(T))}}\right)$, where $N(x)$ is the standard-normal cumulative probability function, see Munk (2004b), p. 163.

[151] Chan et al. (1992), p. 1218.

[152] Calculations of the author. The probability of a negative future short rate in one year is hence 4.6 % and that it is negative in five years is just 2.6 %.

Now, we have a closer look at the term structure of interest rates. With $P(t,T)$ from (3.28) and $R(t,T)$ from (2.2) we can obtain the term structure of interest rates.[153]

$$\begin{aligned} R(t,T) =& -\frac{A(t,T)}{T-t} + \frac{B(t,T)}{T-t} r(t) \\ =& R(\infty) + (r(t) - R(\infty)) \frac{1}{\kappa(T-t)} (1 - \exp(-\kappa(T-t))) \\ &+ \frac{\sigma_r^2}{4\kappa^3(T-t)} (1 - \exp(-\kappa(T-t)))^2 \end{aligned} \tag{3.31}$$

As can be seen, it is affine in $r(t)$ and the Vasicek model hence belongs to the class of affine interest rate models. It can be shown, that the Vasicek model allows for normal, inverted and slightly humped term structures.[154] The corresponding parameters are[155]

- For $r(t) \leq R(\infty) - \frac{\sigma_r^2}{4\kappa^2}$ the term structure of interest rates is monotonically increasing (i.e. normal)
- For $r(t) \geq R(\infty) + \frac{\sigma_r^2}{2\kappa^2}$ it is monotonically decreasing (i.e. inverse)
- For $R(\infty) - \frac{\sigma_r^2}{4\kappa^2} \leq r(t) \leq R(\infty) + \frac{\sigma_r^2}{2\kappa^2}$ it is humped

Next, we discuss how changes in the short rate affect the term structure of interest rates. This can be seen by taking the partial derivative of $R(t,T)$ with respect to $r(t)$

$$\frac{\partial R(t,T)}{\partial r(t)} = \frac{B(t,T)}{T-t} = \frac{1 - \exp(-\kappa(T-t))}{\kappa(T-t)} \tag{3.32}$$

Figure 3.1 shows the effect of changes in r on the term structure of interest rates ($\frac{\partial R}{\partial r}$) as a function of maturity for different values of κ.

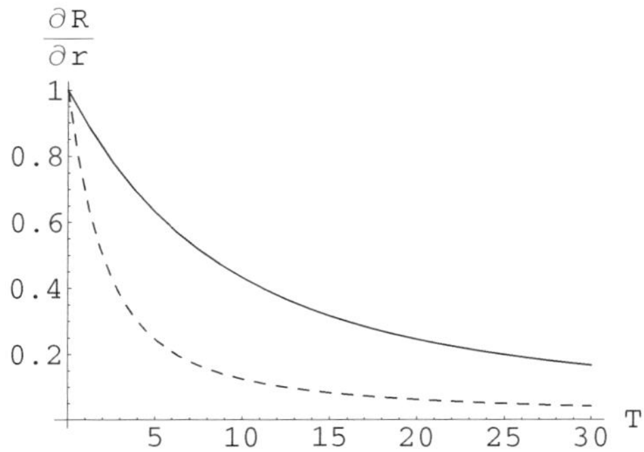

Fig. 3.1. Vasicek model: $\frac{\partial R}{\partial r}$ as a function of maturity.

[153] See Vasicek (1977), p. 186.
[154] See Vasicek (1977), p. 168.
[155] Vasicek (1977), pp. 186–187.

A rise in the short rate has no effect on the infinitely-long rate, because as can be seen from (3.27), it is independent of r. The effect of a change in the short rate on the other interest rates is decreasing in T.[156] The yield on an infinitely-lived bond is fixed at $R(\infty)$ and hence the short rate $r(t)$ moves the entire yield curve.[157]

Now, we are interested in the volatility structure of spot interest rates. In order to obtain it, we first have to derive the spot interest rate dynamics. With dr from (3.16) and $R(t,T)$ from (3.31), Itô's lemma gives

$$\begin{aligned} dR(t,T) &= R_t dt + R_r dr + \frac{1}{2} R_{rr} (dr)^2 \\ &= R_t dt + \frac{B(t,T)}{T-t} dr \\ &= \left(R_t + \frac{B(t,T)}{T-t} \kappa(\theta - r) \right) dt + \left(\frac{B(t,T)}{T-t} \sigma_r \right) dz_1 \\ &= \mu_R(r,t,T) dt + \sigma_R(t,T) dz_1 \end{aligned}$$

The expression $\sigma_R(t,T)$ describes the term structure of volatilities

$$\sigma_R(t,T) = \frac{1 - \exp(-\kappa(T-t))}{\kappa(T-t)} \sigma_r \tag{3.33}$$

Figure 3.2 contains a plot of the volatility structure of spot interest rates

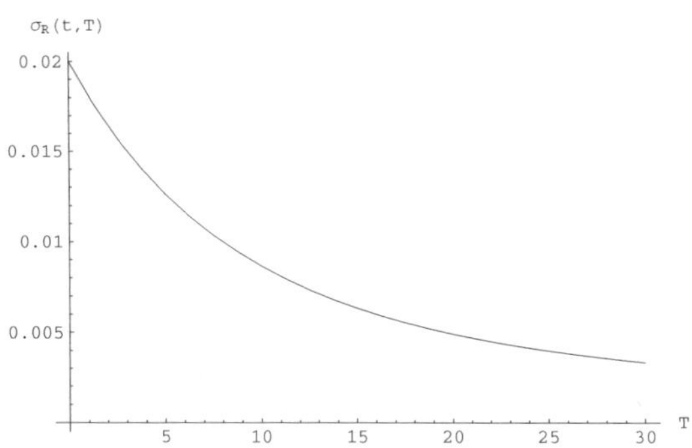

Fig. 3.2. Vasicek model: Volatility structure

It can be seen that the volatility declines when the term to maturity increases, i.e. short term rates are more volatile than long term rates. Empirical studies seem to confirm this volatility structure, but sometimes also humped volatility structures[158] are observed in the marketplace.[159] Humped volatility

[156] Calculate the partial derivative $\frac{\partial^2 R}{\partial r \partial T}$. This expression is negative.
[157] See Holden (2005), p. 55.
[158] The volatility rises at first but then declines.
[159] See Golub/Tilman (2000), p. 89.

structures cannot result from a Vasicek model.[160] The formula also implies constant interest rate volatilities.

A serious drawback of the model in a portfolio selection context is the perfect correlation of the spot rates

$$\begin{aligned}\operatorname{corr}(R(t,T),R(t,\tau)) &= \frac{\operatorname{cov}(R(t,T),R(t,\tau))}{\operatorname{std}(R(t,T))\operatorname{std}(R(t,\tau))}\\ &= \frac{\operatorname{cov}\left(\frac{B(t,T)}{T-t}r(t),\frac{B(t,\tau)}{\tau-t}r(t)\right)}{\operatorname{std}\left(\frac{B(t,T)}{T-t}r(t)\right)\operatorname{std}\left(\frac{B(t,\tau)}{\tau-t}r(t)\right)}\\ &= \frac{\frac{B(t,T)}{T-t}\frac{B(t,\tau)}{\tau-t}\operatorname{var}(r(t))}{\frac{B(t,T)}{T-t}\operatorname{std}(r(t))\frac{B(t,\tau)}{\tau-t}\operatorname{std}(r(t))}\\ &= 1\end{aligned}$$

Zero-coupon bond prices are nonlinear functions of the short rate.[161] Therefore zero-coupon bonds of different maturities are also perfectly (because there is only one source of randomness) but non-linearly correlated, i.e. the correlation coefficient is near but not equal to 1.[162]

A well-known extension to the Vasicek model is due to Hull/White (1990).[163] One drawback of the original Vasicek model, is that it cannot perfectly fit the initial term structure of interest rates.[164] By making the drift of the short rate[165] time-dependent, it is now possible to match any initial term structure of interest rates by construction. In this thesis, we are interested in bond portfolio selection problems. The matching of an initial term structure of interest rates is in our application of no concern, since we assume that the model term structure is equal to the observed initial term structure. Hence, we concentrate on the time-homogeneous models.[166]

3.5 Hull/White (1994)

3.5.1 Introduction

Hull/White (1994) propose a two-factor interest rate model based on Vasicek's term structure model (hereafter, HW2). The two factors are the short rate

---

160 The partial derivative $\frac{\partial \sigma_R(\tau)}{\partial \tau}$ is negative.

161 See Equation (3.28).

162 In the next section, we derive the HW2 model. It allows for non-perfect correlations between interest rates.

163 The so-called extended Vasicek model.

164 Since the model has only a finite number of parameters, these cannot be chosen in such a way as to match the prices of infinitely many bonds.

165 Actually, the mean reversion level.

166 The analysis can easily be extended to time-inhomogeneous models.

r and the mean-reversion level ε. This extension provides for a richer set of possible volatility structures and spot rate correlations. They assume the following dynamics for the factors[167]

$$dr(t) = (\theta + \varepsilon(t) - \kappa_r r(t))dt + \sigma_r dz_1 \tag{3.34}$$

$$d\varepsilon(t) = -\kappa_\varepsilon \varepsilon(t)dt + \sigma_\varepsilon \varrho dz_1 + \sigma_\varepsilon \sqrt{1-\varrho^2}dz_2 \tag{3.35}$$

where θ, σ_r, σ_ε, $\kappa_r > 0$ and $\kappa_\varepsilon > 0$ are constants.[168] The Brownian motions dz_1 and dz_2 are assumed to be uncorrelated and ϱ is the correlation coefficient between the short rate r and the mean-reversion level ε.[169]

3.5.2 Derivation of Zero-Coupon Bond Prices

The HW2 model is also a special case of the general HJM framework. The model is a two-factor interest rate model, i.e. $d = 2$. The forward rate volatilities are assumed to be

$$s_1(t,T) = \frac{e^{(t-T)\kappa_r}\left(\kappa_r - \kappa_\varepsilon\right)\sigma_r - \left(e^{(t-T)\kappa_r} - e^{(t-T)\kappa_\varepsilon}\right)\varrho\sigma_\varepsilon}{\kappa_r - \kappa_\varepsilon} \tag{3.36}$$

$$s_2(t,T) = \frac{\left(-e^{(t-T)\kappa_r} + e^{(t-T)\kappa_\varepsilon}\right)\sqrt{1-\varrho^2}\sigma_\varepsilon}{\kappa_r - \kappa_\varepsilon} \tag{3.37}$$

and the market prices of interest rate risk are constant

$$\lambda_1(t) = \lambda_1 \tag{3.38}$$

$$\lambda_2(t) = \lambda_2 \tag{3.39}$$

The initial forward rate curve is given by

$$f(0,T) = g(0,T) + r(0)e^{-\kappa_r T} + \varepsilon(0)\frac{e^{-\kappa_\varepsilon T} - e^{-\kappa_r T}}{\kappa_r - \kappa_\varepsilon} \tag{3.40}$$

with $g(0,T)$ defined in Equation (A.6) in Appendix A.4.

An application of the general zero-bond pricing equation (3.12) further requires only an expression for the short rate $f(T,T)$. The short rate can be explicitly calculated by application of Equation (3.15)

---

[167] This formulation is equivalent to the original one with correlated Brownian motions. See e.g. Brigo/Mercurio (2001), p. 134

[168] In the original formulation θ is allowed to be a function of time. We restrict ourselves to the special case $\theta(t) = \theta$.

[169] To see this calculate the correlation between $d\varepsilon$ and dr

$$\text{corr}(d\varepsilon, dr) = \frac{\text{cov}(d\varepsilon, dr)}{\text{std}(d\varepsilon)\text{std}(dr)} = \frac{\varrho\sigma_r\sigma_\varepsilon}{\sigma_r\sigma_\varepsilon} = \varrho$$

$$\begin{aligned}f(T,T) =& f(0,T) + \int_0^T s_1(u,T)\left(\int_u^T s_1(u,s)ds - \lambda_1\right)du\\&+ \int_0^T s_2(u,T)\left(\int_u^T s_2(u,s)ds - \lambda_2\right)du\\&+ \int_0^T s_1(u,T)dz_1(u) + \int_0^T s_2(u,T)dz_2(u)\end{aligned}\tag{3.41}$$

We insert (3.40) into (3.41). The resulting expression can be simplified further since it can be shown that

$$\begin{aligned}g(0,T) &+ \int_0^T s_1(u,T)\left(\int_u^T s_1(u,s)ds - \lambda_1\right)du\\&+ \int_0^T s_2(u,T)\left(\int_u^T s_2(u,s)ds - \lambda_2\right)du = \frac{\theta}{\kappa_r}(1-e^{-\kappa_r T})\end{aligned}$$

Hence, we obtain the following formula for the short rate of interest in the HW2 model

$$\begin{aligned}f(T,T) =& \frac{\theta}{\kappa_r}(1-e^{-\kappa_r(T-t)}) + r(t)e^{-\kappa_r(T-t)} + \varepsilon(t)\frac{e^{-\kappa_\varepsilon(T-t)} - e^{-\kappa_r(T-t)}}{\kappa_r - \kappa_\varepsilon}\\&+ \int_t^T \frac{e^{(u-T)\kappa_r}\left(\kappa_r - \kappa_\varepsilon\right)\sigma_r - \left(e^{(u-T)\kappa_r} - e^{(u-T)\kappa_\varepsilon}\right)\varrho\sigma_\varepsilon}{\kappa_r - \kappa_\varepsilon}dz_1(u)\\&+ \int_t^T \frac{\left(-e^{(u-T)\kappa_r} + e^{(u-T)\kappa_\varepsilon}\right)\sqrt{1-\varrho^2}\sigma_\varepsilon}{\kappa_r - \kappa_\varepsilon}dz_2(u)\end{aligned}\tag{3.42}$$

Now, we are in a position to derive the prices of zero-coupon bonds by applying Equation (3.12). With (3.38) and (3.39) we obtain

$$\begin{aligned}P(t,T) =& E_t\left[\exp\left(-\int_t^T f(s,s)ds - \frac{1}{2}\lambda_1^2(T-t) - \frac{1}{2}\lambda_2^2(T-t)\right.\right.\\&\left.\left.+ \int_t^T \lambda_1 dz_1(s) + \int_t^T \lambda_2 dz_2(s)\right)\right]\end{aligned}\tag{3.43}$$

First, we calculate the integral over the short rate

$$
\begin{aligned}
&-\int_t^T f(s,s)ds = \\
&-\int_t^T r(t)e^{-\kappa_r(s-t)}ds - \int_t^T \frac{\theta}{\kappa_r}(1-e^{-\kappa_r(s-t)})ds \\
&-\int_t^T \varepsilon(t)\frac{e^{-\kappa_\varepsilon(s-t)} - e^{-\kappa_r(s-t)}}{\kappa_r - \kappa_\varepsilon}ds \\
&-\int_t^T\int_t^s \frac{e^{(t-s)\kappa_r}\left(\kappa_r-\kappa_\varepsilon\right)\sigma_r - \left(e^{(t-s)\kappa_r} - e^{(t-s)\kappa_\varepsilon}\right)\varrho\sigma_\varepsilon}{\kappa_r-\kappa_\varepsilon}dz_1(u)ds \\
&-\int_t^T\int_t^s \frac{\left(-e^{(t-s)\kappa_r} + e^{(t-s)\kappa_\varepsilon}\right)\sqrt{1-\varrho^2}\sigma_\varepsilon}{\kappa_r-\kappa_\varepsilon}dz_2(u)ds
\end{aligned} \tag{3.44}
$$

The first three integrals in Equation (3.44) are Riemann integrals and can be calculated explicitly. By defining

$$
\begin{aligned}
B_1(t,T) &= \int_t^T e^{-\kappa_r(s-t)}ds \\
&= \frac{1-e^{-\kappa_r(T-t)}}{\kappa_r}
\end{aligned} \tag{3.45}
$$

and

$$
\begin{aligned}
B_2(t,T) &= \int_t^T \frac{e^{-\kappa_\varepsilon(s-t)} - e^{-\kappa_r(s-t)}}{\kappa_r-\kappa_\varepsilon}ds \\
&= \frac{\left(1-e^{(t-T)\kappa_\epsilon}\right)\kappa_r - \left(1-e^{(t-T)\kappa_r}\right)\kappa_\epsilon}{\kappa_r\left(\kappa_r-\kappa_\epsilon\right)\kappa_\epsilon}
\end{aligned} \tag{3.46}
$$

we can write the result as

$$
\begin{aligned}
&-\int_t^T f(s,s)ds = \\
&-r(t)B_1(t,T) - \varepsilon(t)B_2(t,T) - \theta\frac{-1+e^{-\kappa_r(T-t)} + (T-t)\kappa_r}{\kappa_r^2} \\
&-\int_t^T\int_t^s \frac{e^{(t-s)\kappa_r}\left(\kappa_r-\kappa_\varepsilon\right)\sigma_r - \left(e^{(t-s)\kappa_r} - e^{(t-s)\kappa_\varepsilon}\right)\varrho\sigma_\varepsilon}{\kappa_r-\kappa_\varepsilon}dz_1(u)ds \\
&-\int_t^T\int_t^s \frac{\left(-e^{(t-s)\kappa_r} + e^{(t-s)\kappa_\varepsilon}\right)\sqrt{1-\varrho^2}\sigma_\varepsilon}{\kappa_r-\kappa_\varepsilon}dz_2(u)ds
\end{aligned} \tag{3.47}
$$

The last two expressions in Equation (3.47) contain Riemann integrals and stochastic integrals. These expressions can be simplified to expressions containing stochastic integrals only by substitution and integration by parts.[170] We apply the formula derived in Brigo/Mercurio (2001) and obtain the following solution[171]

---

170 For a general integration by parts formula see Oksendal (1992), p. 46.

171 Brigo/Mercurio (2001), p. 150.

$$
\begin{aligned}
-\int_t^T r(s)ds =& -r(t)B_1(t,T) - \varepsilon(t)B_2(t,T) \\
& -\theta\frac{-1+e^{-\kappa_r(T-t)}+(T-t)\kappa_r}{\kappa_r^2} \\
& +\int_t^T \frac{\frac{\left(-1+e^{(s-T)\kappa_\varepsilon}\right)\varrho\sigma_\varepsilon}{\kappa_\varepsilon}+\frac{\left(-1+e^{(s-T)\kappa_r}\right)((\kappa_r-\kappa_\varepsilon)\sigma_r-\varrho\sigma_\varepsilon)}{\kappa_r}}{\kappa_r-\kappa_\varepsilon}dz_1(s) \\
& +\int_t^T \frac{\sqrt{1-\varrho^2}\left(\frac{-1+e^{(s-T)\kappa_\varepsilon}}{\kappa_\varepsilon}-\frac{-1+e^{(s-T)\kappa_r}}{\kappa_r}\right)\sigma_\varepsilon}{\kappa_r-\kappa_\varepsilon}dz_2(s)
\end{aligned}
$$

We define

$$
\eta(s) = \left(\frac{\frac{\left(-1+e^{(s-T)\kappa_\varepsilon}\right)\varrho\sigma_\varepsilon}{\kappa_\varepsilon}+\frac{\left(-1+e^{(s-T)\kappa_r}\right)((\kappa_r-\kappa_\varepsilon)\sigma_r-\varrho\sigma_\varepsilon)}{\kappa_r}}{\kappa_r-\kappa_\varepsilon}+\lambda_1\right)
$$

and

$$
\nu(s) = \left(\frac{\sqrt{1-\varrho^2}\left(\frac{-1+e^{(s-T)\kappa_\varepsilon}}{\kappa_\varepsilon}-\frac{-1+e^{(s-T)\kappa_r}}{\kappa_r}\right)\sigma_\varepsilon}{\kappa_r-\kappa_\varepsilon}+\lambda_2\right)
$$

then the result can be written in a more readable form as

$$
\begin{aligned}
-\int_t^T r(s)ds =& -r(t)B_1(t,T) - \varepsilon(t)B_2(t,T) - \theta\frac{-1+e^{-\kappa_r(T-t)}+(T-t)\kappa_r}{\kappa_r^2} \\
& +\int_t^T \eta(s)dz_1(s) + \int_t^T \nu(s)dz_2(s)
\end{aligned} \tag{3.48}
$$

We insert (3.48) into (3.43) and obtain the following expression for zero-coupon prices

$$
\begin{aligned}
& P(t,T) = \\
& E_t\left[\exp\left(-r(t)B_1(t,T) - \varepsilon(t)B_2(t,T) - \theta\frac{-1+e^{-\kappa_r(T-t)}+(T-t)\kappa_r}{\kappa_r^2}\right.\right. \\
& \left.\left. +\int_t^T \eta(s)dz_1(s) + \int_t^T \nu(s)dz_2 - \frac{1}{2}\lambda_1^2(T-t) - \frac{1}{2}\lambda_2^2(T-t)\right)\right]
\end{aligned}
$$

The only random variables are the two stochastic integrals, hence

$$
\begin{aligned}
P(t,T) =& \exp\left(-r(t)B_1(t,T) - \varepsilon(t)B_2(t,T)\right. \\
& \left. -\theta\frac{-1+e^{-\kappa_r(T-t)}+(T-t)\kappa_r}{\kappa_r^2} - \frac{1}{2}\lambda_1^2(T-t) - \frac{1}{2}\lambda_2^2(T-t)\right) \\
& \times E_t\left[\exp\left(\int_t^T \eta(s)dz_1(s) + \int_t^T \nu(s)dz_2\right)\right]
\end{aligned} \tag{3.49}
$$

Next, we need to calculate the expectations of the exponential of the sum of the two stochastic integrals.[172]

$$
\begin{aligned}
& E_t\left[\exp\left(\int_t^T \eta(s)dz_1(s) + \int_t^T \nu(s)dz_2\right)\right] \\
& = \exp\left(\frac{1}{2}\int_t^T \eta(s)^2 ds + \frac{1}{2}\int_t^T \nu(s)^2 ds\right)
\end{aligned} \tag{3.50}
$$

We insert (3.50) into (3.49) and obtain

$$
\begin{aligned}
P(t,T) = & \exp\Bigg(-r(t)B_1(t,T) - \varepsilon(t)B_2(t,T) - \theta\frac{-1+e^{-\kappa_r(T-t)}+(T-t)\kappa_r}{\kappa_r^2} \\
& -\frac{1}{2}\lambda_1^2(T-t) - \frac{1}{2}\lambda_2^2(T-t) + \frac{1}{2}\int_t^T \eta(s)^2 ds + \frac{1}{2}\int_t^T \nu(s)^2 ds\Bigg)
\end{aligned}
$$

With

$$
\begin{aligned}
A(t,T) = & -\theta\frac{-1+e^{-\kappa_r(T-t)}+(T-t)\kappa_r}{\kappa_r^2} - \frac{1}{2}\lambda_1^2(T-t) - \frac{1}{2}\lambda_2^2(T-t) \\
& + \frac{1}{2}\int_t^T \eta(s)^2 ds + \frac{1}{2}\int_t^T \nu(s)^2 ds
\end{aligned}
$$

and $B_1(t,T)$ and $B_2(t,T)$ from (3.45) and (3.46), this can be written in short form as

$$
P(t,T) = \exp\left(A(t,T) - B_1(t,T)r(t) - B_2(t,T)\varepsilon(t)\right) \tag{3.51}
$$

As can be seen from Equation (3.51), the HW2 model belongs to the affine class as well.

3.5.3 Properties

First, we have a look at the short rate properties. Since this model builds on the Vasicek model, it keeps many of its properties. The short rate is again normally distributed with mean

$$
\begin{aligned}
E_t[r(T)] = & \frac{\theta}{\kappa_r}(1-e^{-\kappa_r(T-t)}) + r(t)e^{-\kappa_r(T-t)} \\
& + \varepsilon(t)\frac{e^{-\kappa_\varepsilon(T-t)} - e^{-\kappa_r(T-t)}}{\kappa_r - \kappa_\varepsilon}
\end{aligned} \tag{3.52}
$$

[172] The sum of the two stochastic integrals is normally distributed. If x is normally distributed, then $E[e^x] = e^{E[x]+\frac{1}{2}\mathrm{var}(x)}$, see Rinne (1997), p. 365. In this case, $E[x] = 0$.

and variance[173]

$$\begin{aligned}
\mathrm{var}_t(r(T)) &= \mathrm{var}_t\left(\int_t^T s_1(u,T)dz_1(u) + \int_t^T s_2(u,T)dz_2(u)\right) \\
&= \int_t^T s_1(u,T)^2 du + \int_t^T s_2(u,T)^2 du \\
&= \int_t^T \left(s_1(u,T)^2 + s_2(u,T)^2\right) du
\end{aligned} \tag{3.53}$$

With $P(t,T)$ from (3.51) and $R(t,T)$ from (2.2), we can obtain the term structure of interest rates

$$R(t,T) = -\frac{A(t,T)}{T-t} + \frac{B_1(t,T)}{T-t} r(t) + \frac{B_2(t,T)}{T-t} \varepsilon(t) \tag{3.54}$$

With this expression for the term structure of interest rates, we are in a position to analyze how changes in the factors influence the spot interest rates. As in the Vasicek model, we can obtain an expression for the interest rate with infinite maturity. The infinitely long rate is

$$\begin{aligned}
R(\infty) &\equiv \lim_{T\to\infty} R(t,T) \\
&= \frac{\kappa_\varepsilon (\theta + \lambda_1 \sigma_r) + \left(\varrho\lambda_1 + \sqrt{1-\varrho^2}\lambda_2\right)\sigma_\varepsilon}{\kappa_r \kappa_\varepsilon} - \frac{\kappa_\varepsilon^2 \sigma_r^2 + 2\varrho\kappa_\varepsilon\sigma_\varepsilon\sigma_r + \sigma_\varepsilon^2}{2\kappa_r^2\kappa_\varepsilon^2}
\end{aligned}$$

It can be seen that it is a constant, i.e. this rate is neither influenced by $r(t)$ nor $\varepsilon(t)$. The effect of changes in the factors on the term structure can be obtained by calculating the respective partial derivatives $\frac{\partial R(t,T)}{\partial r}$ and $\frac{\partial R(t,T)}{\partial \varepsilon}$. The first partial derivative

$$\frac{\partial R(t,T)}{\partial r} = \frac{B_1(t,T)}{T-t} = \frac{1-\exp(-\kappa_r(T-t))}{(T-t)\kappa_r}$$

is equivalent to the Vasicek model.[174] Figure 3.3 shows the effect of changes in $r(t)$ on the shape of the term structure of interest rates for different values of κ_r.

[173] With $s_1(t,T)$ and $s_2(t,T)$ from (3.36) and (3.37).

[174] See Equation (3.32).

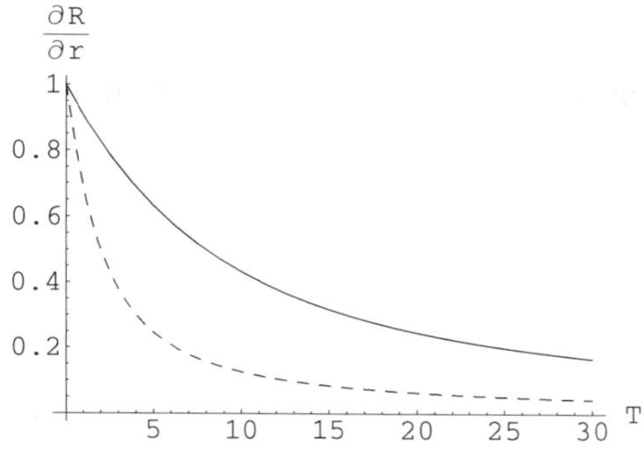

Fig. 3.3. HW2 model: $\frac{\partial R}{\partial r}$ as a function of maturity.

Hence changes in the short rate have a bigger influence on shorter rates than on longer rates. The infinitely long rate is not influenced by changes in the short rate, since its level is independent of both random variables. The second state variable, the mean reversion level, has a different impact on the term structure of interest rates. The second partial derivative is

$$\frac{\partial R(t,T)}{\partial \varepsilon} = \frac{B_2(t,T)}{T-t} = \frac{\left(1 - e^{(t-T)\kappa_\varepsilon}\right)\kappa_r + \left(-1 + e^{(t-T)\kappa_r}\right)\kappa_\varepsilon}{(T-t)\kappa_r\left(\kappa_r - \kappa_\varepsilon\right)\kappa_\varepsilon}$$

As can be seen from Figure 3.4, changes in $\varepsilon(t)$ have no effect on the short rate, the biggest effect on intermediate rates and smaller effect on long rates.

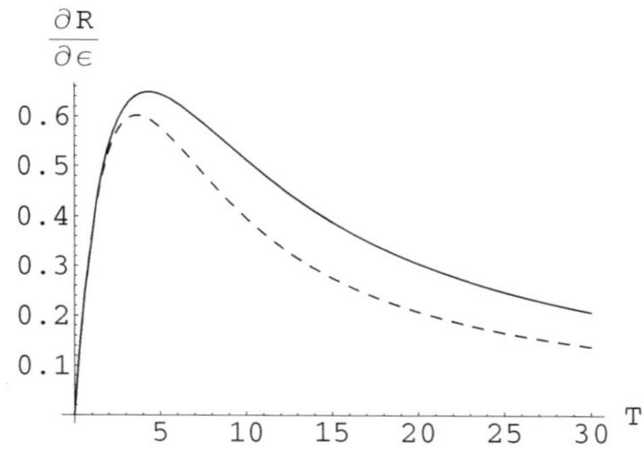

Fig. 3.4. HW2 model: $\frac{\partial R}{\partial \varepsilon}$ as a function of maturity.

Figures 3.3 and 3.4 resemble the factor analysis graphs of principal component analysis of the yield curve.[175] The form of the factor influence is different than empirical observations suggest, but the inclusion of a second (not perfectly correlated) factor clearly has the effect of introducing more realistic

[175] See Litterman/Scheinkman (1991), p. 58.

term structure movements. When short rate changes and mean-reversion level changes are strongly correlated, one can expect unidirectional changes of the whole term structure of interest rates since the partial derivatives have the same sign. Principal component analysis of the term structure of interest rates on the other hand suggests that some factors have widely different impacts on interest rates, i.e. changes in a factor move some spot rates up and some down.

The spot rate volatility structure is also quite different. The Vasicek model allows for a spot rate volatility structure that is decreasing in T. We already pointed out that some empirical studies suggest the existence of humped volatility structures.[176] In contrast to the Vasicek model – as we will show next – the HW2 model allows for these humped volatility structures.

With $R(t,T)$ from (3.54), Itô's lemma can be used to determine the term structure of volatilities[177]

$$\begin{aligned}
dR(t,T) =& R_t dt + R_r dr + R_\varepsilon d\varepsilon + \frac{1}{2}R_{rr}(dr)^2 + \frac{1}{2}R_{\varepsilon\varepsilon}(d\varepsilon)^2 + R_{r\varepsilon}(dr)(d\varepsilon) \\
=& R_t dt + \frac{B_1(t,T)}{T-t}dr + \frac{B_2(t,T)}{T-t}dr \\
=& \left(R_t + \frac{B_1(t,T)}{T-t}(\theta + \varepsilon(t) - \kappa_r r(t)) - \frac{B_2(t,T)}{T-t}\kappa_\varepsilon \varepsilon(t)\right) dt \\
&+ \left(\frac{B_1(t,T)}{T-t}\sigma_r + \frac{B_2(t,T)}{T-t}\sigma_\varepsilon \varrho\right) dz_1 + \frac{B_2(t,T)}{T-t}\sigma_\varepsilon\sqrt{1-\varrho^2}dz_2 \\
=& \mu_R(r,\varepsilon,t,T)dt + \sigma_{1,R}(t,T)dz_1 + \sigma_{2,R}(t,T)dz_2
\end{aligned}$$

With

$$\sigma_{1,R}(t,T) = \frac{B_1(t,T)}{T-t}\sigma_r + \frac{B_2(t,T)}{T-t}\sigma_\varepsilon \varrho \tag{3.55}$$

$$\sigma_{2,R}(t,T) = \frac{B_2(t,T)}{T-t}\sigma_\varepsilon\sqrt{1-\varrho^2} \tag{3.56}$$

it follows that the volatility of spot interest rates is

$$\sigma_R(t,T) = \sqrt{\sigma_{1,R}(t,T)^2 + \sigma_{2,R}(t,T)^2} \tag{3.57}$$

Figure 3.5 provide a graphical representation of the spot rate volatility structure.[178]

---

[176] See Golub/Tilman (2000), p. 89.

[177] dr and $d\varepsilon$ from (3.34) and (3.35).

[178] The following parameter values have been chosen: $\varrho = 0.5, \kappa_r = 0.5, \kappa_\varepsilon = 0.1, \sigma_r = 0.01, \sigma_\varepsilon = 0.04$.

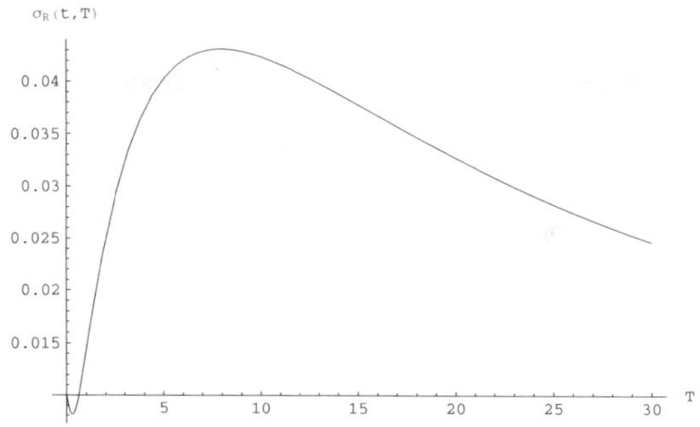

Fig. 3.5. HW2 model: Volatility structure.

Lastly, the introduction of a second factor has an important effect on the correlation structure of spot rates (or zero-coupon bond prices).

$$\begin{aligned}
&\mathrm{corr}(R(t,T), R(t,\tau)) \\
&= \frac{\mathrm{cov}(R(t,T), R(t,\tau))}{\mathrm{std}(R(t,T))\mathrm{std}(R(t,\tau))} \\
&= \frac{\mathrm{cov}\left(\frac{B_1(t,T)}{T-t}r(t) + \frac{B_2(t,T)}{T-t}\varepsilon(t), \frac{B_1(t,\tau)}{\tau-t}r(t) + \frac{B_2(t,\tau)}{\tau-t}\varepsilon(t)\right)}{\mathrm{std}\left(\frac{B_1(t,T)}{T-t}r(t) + \frac{B_2(t,T)}{T-t}\varepsilon(t)\right)\mathrm{std}\left(\frac{B_1(t,\tau)}{\tau-t}r(t) + \frac{B_2(t,\tau)}{\tau-t}\varepsilon(t)\right)}
\end{aligned}$$

This expression is not necessarily equal to 1. Therefore, spot rates (and zero-coupon bond prices) of different maturities are not perfectly correlated in the HW2 model. This is especially interesting in a portfolio context, as we will see in the next chapter.

The original formulation of the model assumed that θ is not constant but a deterministic a function of time.[179] With this extension, the model is able to match any particular initial term structure. Furthermore, the volatility functions σ_r and σ_ε can be made deterministic functions of time as well and hence the model could match any volatility structure. But this is generally problematic since doing so results in unrealistic deterministic movements of the volatility structure over time.[180]

3.6 Summary and Conclusion

In this chapter we derived the general HJM framework using the stochastic discount factor approach. Furthermore, we derived and analyzed two special cases, the one-factor Vasicek (1977) model and the two-factor HW2 model.

[179] See Hull/White (1996), p. 334.
[180] See Hull/White (1996), p. 369.

From a portfolio selection perspective, the HW2 model has the advantage of capturing the real-world interest rate correlations better. In the Vasicek model, all interest rates are perfectly correlated. Also the possible term structure movements in the HW2 model are more realistic than in the Vasicek model.

In Chapters 4 and 5 we incorporate the Vasicek (1977) and the HW2 models into bond portfolio optimization problems.

4

Static Bond Portfolio Optimization

4.1 Introduction

This chapter describes static bond portfolio optimization based on the mean-variance framework of Markowitz (1952). This theoretical approach for bond portfolio selection was first introduced by Wilhelm (1992). We show why and how the Markowitz model has to be adapted in order to be useful for the selection of bond portfolios. After deriving the adjusted portfolio selection problem, we apply it to the Vasicek and the HW2 model, we presented in Chapter 3.[181] Furthermore, numerical examples highlight potential problems and inner workings of the model. In the last part of the chapter, this theoretical bond portfolio selection model is compared – by means of numerical examples – to well-known active and passive portfolio selection methods used in practice.[182]

4.2 Static Bond Portfolio Selection in Theory

4.2.1 A Short Review of Modern Portfolio Theory

Modern portfolio theory[183] – introduced by Markowitz in 1952[184] – is the "cornerstone of modern asset management".[185] It is a static model in the sense that the investor is assumed to construct a portfolio today and to sell it at a later date (the investment horizon). Between today and the investment horizon, there is no portfolio re-balancing.[186] Realistically portfolio choice

[181] Wilhelm (1992) applies this framework to the term structure model by Cox/Ingersoll/Ross (1985).

[182] A definition of active and passive strategies follows later.

[183] A well-known textbook covering most aspects of modern portfolio theory is Elton et al. (2003).

[184] Markowitz (1952).

[185] Scherer (2002), p. 1.

should be analyzed in a multi-period setting. But the problem then becomes far more complicated and less tractable. Theoretically, a multi-period portfolio selection problem reduces to a sequence of single-period problems (so-called myopic portfolio choice) only if[187]

- returns are independent and identically-distributed random variables
- the investor has a utility function with constant relative risk aversion, i.e. RRA is independent of wealth.

A second assumption of the model is that the investor cares only about expected terminal portfolio wealth (or portfolio return) and variance of terminal portfolio wealth (or variance of portfolio return). Economic theory on the other hand claims, that a rational investor maximizes his expected utility of terminal wealth (or consumption).[188] It has been shown, that the investor's expected utility of terminal wealth is a function of the mean and the variance of terminal wealth only if the preferences of the investor are governed by a quadratic utility function or asset prices are multi-variate normally distributed.[189]

With these assumptions, the problem of the investor is to minimize the portfolio variance given an expected terminal wealth and a budget constraint.

$$\begin{aligned} &\min_N \operatorname{var}(W_T) \\ \text{s.t. } &E[W_T] = \bar{W_T} \\ &\sum_{i=1}^{n} N_i P_i = W_0 \end{aligned} \tag{4.1}$$

where W_0 is the initial and W_T is terminal wealth, N_i is the quantity of asset i in the portfolio, P_i is the price of asset i, E is the expectation and var is the variance operator. The desired expected terminal wealth of the investor is denoted by $\bar{W_T}$. Other constraints could be imposed as well, e.g. short-sale constraints.[190] But closed-form solutions are only available for the unrestricted case.[191] For all other cases, there exists a fast algorithm – the critical line method – for computing the efficient frontier in a parameterized form.[192] For a more recent exposition of the critical line method, that includes

[186] In the real world, the investor will most likely rebalance his portfolio between these two dates. But this is not reflected in the model.

[187] See Mossin (1968), p. 228.

[188] See for example Theorem 3 in Ingersoll (1987), p. 31.

[189] See Ingersoll (1987), p. 96.

[190] $N_i \geq 0 \; \forall \, i = 1, \ldots, n$

[191] The general solution to the mean-variance portfolio selection problem is well known and will not be given here since – as will be shown in the next section – the problem has first to be adapted to portfolios of bonds.

[192] The critical line method was developed by Markowitz and is described in Markowitz (1956) and Markowitz (1959).

modern results from operations research, see Rudolf (1994), Markowitz/Todd (2000) and Mertens (2006).

4.2.2 Application to Bond Portfolios

The formulation of the portfolio selection problem in (4.1) applies in principle to all tradeable assets. But an application to the selection of bonds creates difficulties because the differences between stocks and bonds have to be taken into account since they affect the calculation of the terminal wealth W_T.

One of the biggest differences between stocks and bonds is the finite maturity of the latter asset class. This difference introduces an important problem to the above formulation of the optimization program. Bonds with a maturity less than the investment horizon won't exist at the investment horizon anymore. Hence, we need an assumption about the reinvestment opportunities for cash flows – face value or coupons – received before the investment horizon T.[193] This reinvestment assumption must meet a certain requirement: it has to be possible with the information available at the time of the cash flows to decide on an optimal policy. Hence we can't assume that we rebalance the portfolio every time a cash flows occurs, since this is a classical dynamic programming problem and in such a problem the decision to be taken at time t has to anticipate later decisions.[194] For the static mean variance framework, we need a reinvestment assumption that doesn't need to anticipate future portfolio rebalancing decisions. Wilhelm (1992) assumed – in accordance with the standard duration strategy – that all cash flows received at time $t < T$ shall be invested at the current spot interest rate $R(t,T)$ until the investment horizon T.[195, 196] In order to execute this strategy optimally at time t no anticipation of future decisions at times t_i with $t < t_i < T$ is necessary. Other reinvestment assumptions are possible but this one seems to be a reasonable approximation to reality.[197]

In our analysis, we want to restrict our attention to portfolios of zero-coupon bonds. This is no limitation since we can think of coupon bonds simply as portfolios of zero-coupon bonds.[198] In this regard zero-coupon bonds can

[193] See Wilhelm (1992), p. 216.

[194] See Wilhelm (1992), p. 216.

[195] See Wilhelm (1992), pp. 216–217.

[196] We hence assume that suitable zero-coupon bonds exist on all cash-flow dates.

[197] In Chapter 5 we will consider a dynamic portfolio selection model where the portfolio can be rebalanced continuously.

[198] Example: A 2 year 5% coupon bond with a face value of 100 can be regarded as a portfolio of a one year zero-coupon bond with face value 5 and a two year zero-coupon bond with face value 105.

be thought of as the basic portfolio building blocks.[199] This approach can of course be easily generalized to coupon bonds.[200]

Let the investment universe[201] consist of zero-coupon bonds of different maturities. The longest maturity of all zero-coupon bonds is denoted by τ.[202] There exists one bond for each maturity date $1, 2, ..\tau - 1, \tau$.

At time $t = 0$ the investor allocates his initial wealth W_0 to the τ zero-coupon bonds.

$$W_0 = \sum_{t=1}^{\tau} N_t P(0, t)$$

where N_t denotes the purchased quantity of the zero-coupon bond with maturity date t and current price $P(0, t)$. The τ zero-coupon bonds can be divided into one riskless and $\tau - 1$ risky assets. An investment in the zero-coupon bond with maturity T is riskless. All other zero-coupon bonds are risky investments and are – for notational convenience – combined in the holdings vector $\hat{N}$ and the price vector $\hat{P}_0$

$$W_0 = \hat{N}'\hat{P}_0 + N_T P(0, T) \tag{4.2}$$

with[203]

$$\begin{aligned} \hat{N}' &= (N_1, \ldots, N_{T-1}, N_{T+1}, \ldots, N_\tau) \\ \hat{P_0}' &= (P(0,1), \ldots, P(0, T-1), P(0, T+1), \ldots, P(0, \tau)) \end{aligned}$$

N_T denotes the number of T-maturity bonds the investor purchases at time zero. Next, we derive the terminal wealth of the investor. An investment of $N_T P(0, T)$ in the riskless T-maturity bond at time zero grows to N_T at time T. Holdings of zero-coupon bonds with a maturity greater than the investment horizon are at time T simply worth the sum of the (arbitrage-free) prices of the individual zero-coupon bonds. Holdings of zero-coupon bonds with a maturity less than T are more difficult to value at T. The face value of these zero-coupon bonds is reinvested at time $t < T$ at the current spot rate $R(t, T)$ until the investment horizon. Hence, the terminal wealth is[204]

$$W_T = \sum_{t=1}^{T-1} N_t \exp((T - t)R(t, T)) + N_T + \sum_{t=T+1}^{\tau} N_t \exp(-(t - T)R(T, t))$$

With $R(t, T)$ from (2.2) it follows that

199 Furthermore, this approach is more akin to the real world approach. In practice one rarely observes “government bond picking” but the portfolio managers position themselves appropriately on the yield curve.

200 For a formulation with coupon bonds, the interested reader is referred to Wilhelm (1992).

201 Or investment opportunity set.

202 Generally τ will be greater than T but this is not required.

203 The symbol $'$ denotes transpose.

204 See Wilhelm (1992), p. 217.

$$W_T = \sum_{t=1}^{T-1} N_t \frac{1}{P(t,T)} + N_T + \sum_{t=T+1}^{\tau} N_t P(T,t)$$

Let

$$\hat{P_T} = \left(\frac{1}{P(1,T)}, \ldots, \frac{1}{P(T-1,T)}, P(T,T+1), \ldots, P(T,\tau) \right)'$$

then the terminal wealth can be written in vector notation as follows

$$W_T = \hat{N}'\hat{P_T} + N_T \tag{4.3}$$

In the mean-variance framework, the investor cares only about the expected value and variance of terminal wealth. We apply the expectation operator to Equation (4.3) and obtain[205]

$$\begin{aligned} E[W_T] &= \sum_{t=1}^{T-1} N_t E\left[\frac{1}{P(t,T)}\right] + N_T + \sum_{t=T+1}^{\tau} N_t E\left[P(T,t)\right] \\ &= \hat{N}'E\left[\hat{P_T}\right] + N_T \end{aligned} \tag{4.4}$$

and the variance of terminal wealth is

$$\begin{aligned} \mathrm{var}(W_T) =& \sum_{t=1}^{T-1}\sum_{s=1}^{T-1} N_t N_s \mathrm{cov}\left(\frac{1}{P(t,T)}, \frac{1}{P(s,T)}\right) \\ &+ \sum_{t=T+1}^{\tau}\sum_{s=T+1}^{\tau} N_t N_s \mathrm{cov}\left(P(T,t), P(T,s)\right) \\ &+ 2\sum_{t=1}^{T-1}\sum_{s=T+1}^{\tau} N_t N_s \mathrm{cov}\left(\frac{1}{P(t,T)}, P(T,s)\right) \\ =& \hat{N}'C\hat{N} \end{aligned} \tag{4.5}$$

where C is the covariance matrix. Entries of C are denoted by $\sigma_{i,j}$ with $i,j = 1,\ldots,\tau-1$. The matrix C is defined as follows[206]

$$\sigma_{i,j} = \begin{cases} \mathrm{cov}\left(\frac{1}{P(i,T)}, \frac{1}{P(j,T)}\right) & \text{for } i = 1,2,\ldots,T-1 \\ & \quad\; j = 1,2,\ldots,T-1 \\ \mathrm{cov}\left(\frac{1}{P(i,T)}, P(T,j+1)\right) & \text{for } i = 1,2,\ldots,T-1 \\ & \quad\; j = T,T+1,\ldots,\tau-1 \\ \mathrm{cov}\left(P(T,i+1), \frac{1}{P(j,T)}\right) & \text{for } i = T,T+1,\ldots,\tau-1 \\ & \quad\; j = 1,2,\ldots,T-1 \\ \mathrm{cov}\left(P(T,i+1), P(T,j+1)\right) & \text{for } i = T,T+1,\ldots,\tau-1 \\ & \quad\; j = T,T+1,\ldots,\tau-1 \end{cases} \tag{4.6}$$

[205] See Wilhelm (1992), p. 217.

[206] We assume that $\tau > T$.

For a maximum maturity date of $\tau = 5$ and an investment horizon of $T = 3$ the covariance matrix C would be given by

$$\begin{pmatrix} \mathrm{var}\left(\frac{1}{P(1,3)}\right) & \mathrm{cov}\left(\frac{1}{P(1,3)}, \frac{1}{P(2,3)}\right) & \mathrm{cov}\left(\frac{1}{P(1,3)}, P(3,4)\right) & \mathrm{cov}\left(\frac{1}{P(1,3)}, P(3,5)\right) \\ \cdots & \mathrm{var}\left(\frac{1}{P(2,3)}\right) & \mathrm{cov}\left(\frac{1}{P(2,3)}, P(3,4)\right) & \mathrm{cov}\left(\frac{1}{P(2,3)}, P(3,5)\right) \\ \cdots & \cdots & \mathrm{var}\left(P(3,4)\right) & \mathrm{cov}\left(P(3,4), P(3,5)\right) \\ \cdots & \cdots & \cdots & \mathrm{var}\left(P(3,5)\right) \end{pmatrix}$$

With the initial wealth from (4.2), the expected terminal wealth from (4.4) and the variance of terminal wealth from (4.5) the bond portfolio selection problem can be stated as follows[207]

$$\min_{\hat{N}} \frac{1}{2}\hat{N}'C\hat{N} \tag{4.7}$$

$$\text{s.t. } \hat{N}'E\left[\hat{P}_T\right] + N_T = \bar{W}_T \tag{4.8}$$

$$\hat{N}'\hat{P}_0 + N_T P(0,T) = W_0 \tag{4.9}$$

We can combine Equations (4.8) and (4.9) and obtain

$$\begin{aligned} &\min_{\hat{N}} \frac{1}{2}\hat{N}'C\hat{N} \\ &\text{s.t. } \underbrace{\frac{W_0}{P(0,T)}}_{(*)} + \hat{N}' \underbrace{\left(E\left[\hat{P}_T\right] - \frac{1}{P(0,T)}\hat{P}_0\right)}_{(**)} = \bar{W}_T \end{aligned} \tag{4.10}$$

where $(*)$ denotes the riskfree compounded initial wealth and $(**)$ denotes the vector of risk premia. This problem looks very much identical to the equity formulation[208], but the bond specific adjustments are "hidden" in the vectors $\hat{P}_0$ and $E\left[\hat{P}_T\right]$ as well as in the covariance matrix C. In order to calculate mean-variance-efficient bond portfolios, we therefore need the following input data[209]

[207] As usual we minimize half the variance.

[208] This was intended.

[209] See Wilhelm (1992), p. 217.

Table 4.1. Bond portfolio optimization: Input parameters.

Expression	Parameter	Description
$E[P(T,t)]$	$t = T+1,\ldots,\tau$	expected discount factors at time T for all maturities greater than T
$E\left[\frac{1}{P(t,T)}\right]$	$t = 1,2,\ldots,T-1$	expected accrual factors from t to T
$\text{cov}\left(\frac{1}{P(t,T)}, \frac{1}{P(s,T)}\right)$	$s,t = 1,2,\ldots,T-1$	covariances between different accrual factors
$\text{cov}\left(P(T,t), P(T,s)\right)$	$s,t = T+1,\ldots,\tau$	covariances between discount factors at T
$\text{cov}\left(\frac{1}{P(t,T)}, P(T,s)\right)$	$t = 1,2,\ldots,T-1$ and $s = T+1,\ldots,\tau$	covariances between accrual factors and discount factors

Problem (4.10) is a quadratic optimization problem with one equality constraints. It can be solved by differentiating the LAGRANGE function with respect to $\hat{N}$.[210]

$$\frac{d}{d\hat{N}}\left(\frac{1}{2}\hat{N}'C\hat{N} + \lambda\left(\bar{W_T} - \hat{N}'\left(E\left[\hat{P_T}\right] - \frac{1}{P(0,T)}\hat{P_0}\right) - \frac{W_0}{P(0,T)}\right)\right) = \underline{0}$$

where $\underline{0}$ is a vector of zeros. We obtain the following solution for the zero-coupon bond holdings vector

$$\hat{N} = \lambda \underbrace{\left(C^{-1}E\left[\hat{P_T}\right] - \frac{1}{P(0,T)}C^{-1}\hat{P_0}\right)}_{\text{Tobin fund } y} = \lambda y \tag{4.11}$$

The efficient portfolio of risky assets is a multiple of the so-called Tobin fund y. Next, we insert the optimal solution $\hat{N}$ in the equality constraint and solve for λ. This yields the optimal value for the individual parameter λ as a function of initial and expected future wealth.

$$\lambda = \frac{\bar{W_T} - \frac{W_0}{P(0,T)}}{\left(E\left[\hat{P_T}\right] - \frac{1}{P(0,T)}\hat{P_0}\right)' C^{-1}\left(E\left[\hat{P_T}\right] - \frac{1}{P(0,T)}\hat{P_0}\right)} \tag{4.12}$$

With the solution for λ and $\hat{N}$, the bond portfolio selection problem is solved. If $\bar{W_T} = \frac{W_0}{P(0,T)}$, i.e. the desired expected wealth in T is equal to the riskless compounded wealth, then from Equation (4.12) it can be seen that $\lambda = 0$ and the entire wealth is invested in the T-maturity zero-coupon bond. The entire initial wealth is invested in the risky assets if $\lambda = \frac{W_0}{\hat{P_0} \cdot y}$, the resulting portfolio is called the tangency portfolio.

[210] See Bronstein et al. (1999), p. 396.

$$N_{tan} = \frac{W_0}{\hat{P}_0 \cdot y} y \tag{4.13}$$

The problem of obtaining the necessary parameters is discussed in the next section.

4.2.3 Obtaining the Parameters

In equity portfolio selection, the necessary input parameters (expected returns and covariances of returns) can be obtained by analyzing the time series of the assets. One approach is to impose no structure on the expected returns and covariances of returns and use sample mean and sample covariances as inputs.[211] At the other end of the spectrum are single- or multi-index models for the asset returns.[212] A compromise between these two approaches are so-called (James/Stein) shrinkage estimation techniques.[213]

In bond portfolio optimization, the problem of obtaining the necessary input parameters is more difficult as will become clearer from the following comments. We examine every parameter group from Table 4.1 concerning possible estimation procedures.

The expected zero-coupon bond prices at the investment horizon $E[P(T,t)]$ must be known. Using simple time series analysis is not recommended, since bonds have finite maturities and promise to pay the face value at maturity, so the probability distribution depends on time to maturity.[214] Consider a 5-year zero-coupon bond at issuance. The price at the end of the year is random. Consider the same bond 4 years later. Then it has become a 1-year zero-coupon bond and so the price at the end of the year is non-random, because it pays the face value at maturity. The return distribution is hence time dependent. A T-year zero-coupon bond today becomes a $(T-1)$-year zero-coupon bond in one year's time, hence using the whole time series for expected return or variance estimation is not advised, since these price observations are not for the same asset.[215] The problem can be mitigated by analyzing artificial time series for so-called constant maturity bonds.[216] Since knowledge of the expected zero-bond prices (discount curve) at the investment horizon is equivalent to the knowledge of the expected term structure of interest rates, one could also derive these input parameters by estimating the term structure

---

[211] This is proposed in the classic book by Markowitz (1959).

[212] Because they impose a fixed structure on the asset returns. This approach is explored in Chapter 7 and 8 in Elton et al. (2003).

[213] This approach was suggested by James/Stein (1961). Its application to portfolio selection has been explored in several papers, e.g. Jorion (1986) and Ledoit/Wolfe (2004).

[214] See Wilhelm (1992), p. 213.

[215] At the beginning for a T-year zero-coupon bond and at the end for 1-year zero-coupon bond.

[216] See Elton et al. (2003), p. 545.

of interest rates for each day in a certain historical period and estimating the sample distribution parameters. Any of the classical term structure theories presented in Chapter 2.5 could as well be used to obtain these parameters.[217] Furthermore, dynamic term structure models can be employed. The expected accrual factors $E\left[\frac{1}{P(t,T)}\right]$ for every date before the investment horizon, could be obtained in the same way.

The covariances $\operatorname{cov}(P(T,t), P(T,s))$ can be obtained by analyzing constant maturity bonds as described above or using dynamic term structure theories. Classical term structure theories yield no usable information.

The problem becomes more difficult when we consider $\operatorname{cov}\left(\frac{1}{P(t,T)}, \frac{1}{P(s,T)}\right)$ and $\operatorname{cov}\left(\frac{1}{P(t,T)}, P(T,s)\right)$. Here the covariance between functions of spot interest rates (i.e. accrual and discount factors) for different dates or with different maturity must be derived. Term structure estimation gives just a snapshot of the current term structure and no indication of how the term structure moves over time. So it is impossible to get information about the co-movements between term structures at different dates. Only dynamic term structure models deliver these kind of information.

We can conclude that the input parameters for the bond portfolio optimization problem in (4.10) can only be obtained by employing dynamic term structure models and hence imposing a fixed structure on the bond market. If the investment horizon is the next possible date, then there exist other possibilities for obtaining the parameters and dynamic term structure models are not the only choice.[218]

In theory, every dynamic term structure model could be used for obtaining the bond portfolio selection parameters. In practice, there is a trade-off between realism and analytical tractability.

In Chapter 3 we introduced two special cases of the general HJM framework, the Vasicek model and the HW2 model. These two models are members of the Gaussian affine class – first studied by Langetieg (1980).[219] In our analysis we want to restrict our attention to these two models.[220]

In a Gaussian affine model, zero-coupon bond prices are of the form[221]

---

[217] See Elton et al. (2003), p. 545.

[218] Elton et al. (2003) suggests classical term structure theories and single- or multi-index models, see Elton et al. (2003), pp. 540–546.

[219] The affine models were proposed by Duffie/Kan (1996).

[220] The advantage of the above models is their analytical tractability. If no analytical solution for zero-coupon bond prices exists, then the parameters cannot be calculated explicitly and we might have to resort to approximations to the normal distribution. Wilhelm (1992) assumed that the term structure follows the Cox/Ingersoll/Ross (1985) model. In this model the short rate is chi-squared distributed. He then used an approximation to the normal distribution in order to calculate the bond portfolio selection parameters.

[221] Cairns (2004), p. 103.

$$
\begin{aligned}
P(t,T) &= \exp(A(t,T) - \sum_{j=1}^{k} B_j(t,T)x_j(t)) \\
&= \exp(A(t,T) - B(t,T)'x(t))
\end{aligned} \tag{4.14}
$$

where $x(t)$ is a k-dimensional vector of state variables. Furthermore it is assumed that the state variables are k-dimensionally normally distributed with mean $E[x(t)]$ and covariance matrix $\mathrm{COV}(x(t), x(t))$.[222] Since the short rate is a linear combination of these state variables, the short rate is normally distributed as well.

With the equation for zero-coupon bond prices in (4.14) and the assumption of k-dimensionally normally distributed state variables, we obtain the following results for the bond portfolio selection parameters.[223] The expected discount factors (zero-coupon bond prices at time t with maturity T) are

$$
\begin{aligned}
E\left[P(t,T)\right] =& E\left[\exp(A(t,T) - B(t,T)'x(t))\right] \\
=& \exp\left(A(t,T) - B(t,T)'E[x(t)]\right. \\
& \left. + \frac{1}{2}B(t,T)'\mathrm{COV}(x(t), x(t)')B(t,T)\right)
\end{aligned} \tag{4.15}
$$

The expected accrual factors from t to T can be given by

$$
\begin{aligned}
E\left[\frac{1}{P(t,T)}\right] =& E\left[\exp(-A(t,T) + B(t,T)'x(t))\right] \\
=& \exp\left(-A(t,T) + B(t,T)'E[x(t)]\right. \\
& \left. + \frac{1}{2}B(t,T)'\mathrm{COV}(x(t), x(t)')B(t,T)\right)
\end{aligned} \tag{4.16}
$$

The covariances between discount factors at time t for maturities T and τ are

$$
\begin{aligned}
\mathrm{cov}\left(P(t,T), P(t,\tau)\right) =& E[P(t,T)]E[P(t,\tau)] \\
& \times \left(e^{B(t,T)'\mathrm{COV}(x(t), x(t)')B(t,\tau)} - 1\right)
\end{aligned} \tag{4.17}
$$

The covariances between accrual factors for maturity T at times t and τ can be calculated as follows

$$
\begin{aligned}
\mathrm{cov}\left(\frac{1}{P(t,T)}, \frac{1}{P(\tau,T)}\right) =& E\left[\frac{1}{P(t,T)}\right] E\left[\frac{1}{P(\tau,T)}\right] \\
& \times \left(e^{B(t,T)'\mathrm{COV}(x(t), x(\tau)')B(\tau,T)} - 1\right)
\end{aligned} \tag{4.18}
$$

The covariances between accrual factors at time t with maturity T and discount factors at time T with maturity τ are given next

[222] See Langetieg (1980), p. 82.

[223] See Table 4.1. If x is normally distributed, then $E[e^x] = e^{E[x] + \frac{1}{2}\mathrm{var}(x)}$ and $\mathrm{var}(e^x) = E[e^x]^2(e^{\mathrm{var}(x)} - 1)$, see Rinne (1997), p. 365

$$\operatorname{cov}\left(\frac{1}{P(t,T)}, P(T,\tau)\right) = E\left[\frac{1}{P(t,T)}\right] E\left[P(T,\tau)\right] \times \left(e^{-B(t,T)'\mathrm{COV}(x(t),x(T)')B(T,\tau)} - 1\right) \quad (4.19)$$

In order to calculate the above expressions, we only need the term structure model dependent functions $A(t,T)$ and $B(t,T)$ and the distributional parameters of the state variables. For the Vasicek and the HW2 model, these expression have already been derived in Chapter 3.

In Section 4.2.4 and 4.2.5 we analyze static bond portfolio optimization inside the Vasicek model and the HW2 model by means of numerical examples.

4.2.4 One-Factor Vasicek (1977) Model

We assume the existence of a bond market where zero-coupon bonds of maturities $1, \ldots, \tau = 10$ trade. This provides for a large but still manageable set of bonds. The investment horizon of the investor is denoted by T with $1 \leq T \leq \tau$. The investor can therefore choose between nine risky zero-coupon bonds and one riskless zero-coupon bond (the T-maturity bond).

We consider two investors who differ in their investment horizon: short-term investment horizon ($T = 1$) and long-term investment horizon ($T = 5$). The difference is of importance, since in the short-term case the portfolio value at the investment horizon is solely determined by the term structure at the investment horizon. In the long-term case, on the other hand, it is also influenced by the term structures before the investment horizon.

We introduced the Vasicek (1977) model already in Chapter 3.4. The short rate is the only state variable and it is normally distributed. Hence, the general portfolio optimization input parameters from (4.15) to (4.19) can be specified as follows

$$E\left[\frac{1}{P(t,T)}\right] = \exp\left(-A(t,T) + B(t,T)E[r(t)] + \frac{1}{2}B(t,T)^2 \mathrm{var}(r(t))\right)$$

$$E\left[P(t,T)\right] = \exp\left(A(t,T) - B(t,T)E[r(t)] + \frac{1}{2}B(t,T)^2 \mathrm{var}(r(t))\right)$$

$$\operatorname{cov}\left(\frac{1}{P(t,T)}, \frac{1}{P(\tau,T)}\right) = E\left[\frac{1}{P(t,T)}\right] E\left[\frac{1}{P(\tau,T)}\right] \left(e^{B(t,T)B(\tau,T)\mathrm{cov}(r(t),r(\tau))} - 1\right)$$

$$\operatorname{cov}\left(P(t,T), P(t,\tau)\right) = E[P(t,T)]E[P(t,\tau)] \left(e^{B(t,T)B(t,\tau)\mathrm{var}(r(t))} - 1\right)$$

$$\text{cov}\left(\frac{1}{P(t,T)}, P(T,\tau)\right) = E\left[\frac{1}{P(t,T)}\right] E\left[P(T,\tau)\right]\left(e^{-B(t,T)B(T,\tau)\text{cov}(r(t),r(T))} - 1\right)$$

with $A(t,T)$ and $B(t,T)$ defined in Equations (3.26) and (3.25). In order to calculate the above expressions, we additionally need the expected value and the variance of the short rate and the covariances between short rates at different times. These have already been calculated in (3.29) and (3.30)

$$\begin{aligned} E_t[r(T)] &= r(t)\exp(-\kappa(T-t)) + \theta(1-\exp(-\kappa(T-t))) \\ \text{var}_t(r(T)) &= \sigma_r^2\left(\frac{1-\exp(-2\kappa(T-t))}{2\kappa}\right) \\ \text{cov}_t(r(T), r(\tau)) &= \sigma_r^2\left(\frac{1-\exp(-2\kappa(\min(T,\tau)-t))}{2\kappa}\right) \end{aligned}$$

The term structure parameters – the current level of the short rate $r(0)$, the mean reversion speed κ, the short rate volatility σ_r, the mean reversion level θ and the market price of interest rate risk λ – are normally chosen in such a way to match as closely as possible the current term structure of interest rates (perhaps subject to some economically sensible constraints for certain parameters).[224] For our numerical example, we assume the following values for the parameters[225]

Table 4.2. Vasicek model: Parameter values for numerical example.

Parameter	Value
r(0)	0.0258
θ	0.024
κ	0.1668
σ_r	0.0153
λ	0.2126

As can be seen from Figure 4.1 these parameters describe a normal term structure, i.e. monotonically increasing spot interest rates.

Short-term investment horizon

The short rate is the only systematic risk factor in the Vasicek model.[226] Therefore, we first give the distributional properties of the risk factor and

[224] See Munk (2004b), p. 181.

[225] These values give a close approximation of the German government term structure in January 2006.

[226] There are no idiosyncratic risk factors.

then deduce the distributional parameters of the assets (the zero-coupon bond prices). The current value of the short rate determines the current shape of the term structure of interest rates[227], so the shape of the term structure at the investment horizon is also determined by the future value of the short rate only. The short rate at time $T = 1$ is normally distributed with mean[228] $E[r(1)] = 0.0255235$ and standard deviation[229] $\text{std}(r(1)) = 0.0141084$. Given the distribution of the short rate we can determine the distribution of all interest rates at time T.[230] Spot interest rates are normally distributed with mean[231]

$$E_0[R(t,T)] = -\frac{A(t,T)}{T-t} + \frac{B(t,T)}{T-t} E_0[r(t)]$$

and variance

$$\text{var}_0(R(t,T)) = \left(\frac{B(t,T)}{T-t}\right)^2 \text{var}_0(r(t))$$

From Figure 4.2 we observe that short term rates are more volatile than longer rates.[232]

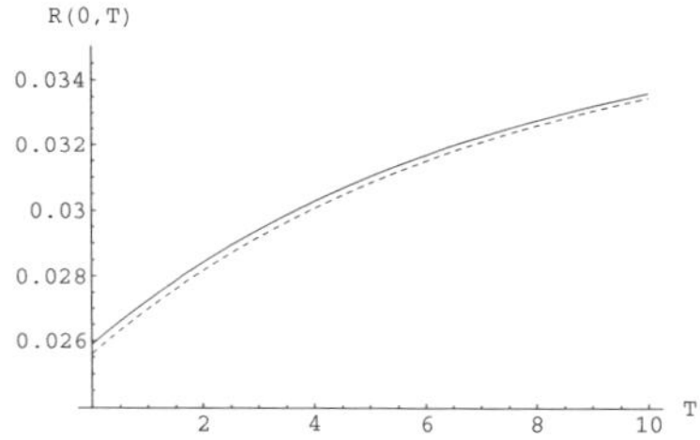

Fig. 4.1. Vasicek: Term structure

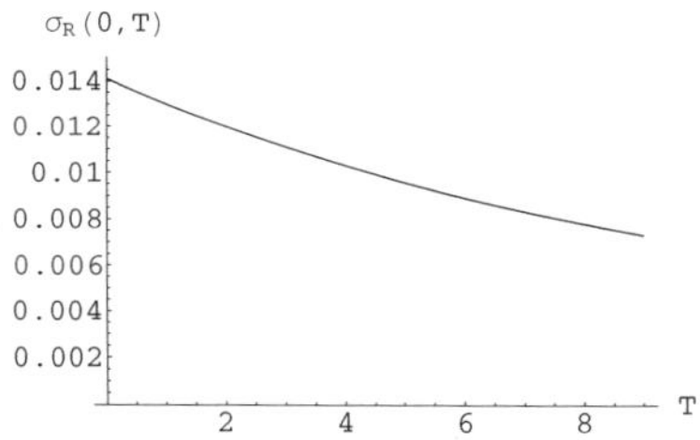

Fig. 4.2. Vasicek: Volatility structure.

An important feature of any term structure model is the ability to simultaneously match the current term structure of interest rates and the covariance structure of interest rates. Matching the current term structure would be possible by introducing a time-dependent drift as in the extended Vasicek model.[233] But matching the covariance structure is impossible. Spot interest rates of different maturities are functions of the short rate only. In the Vasicek

[227] Given the constant parameters.

[228] See Equation (3.29).

[229] See Equation (3.30).

[230] See Equation (3.31) for an expression of $R(t,T)$ in terms of $r(t)$.

[231] Apply the expectations and variance operator to Equation (3.31).

[232] This is empirically observable in the fixed income markets, but humped shapes – which cannot result from a Vasicek model – are also common, see Golub/Tilman (2000), p. 89.

[233] Hull/White (1990).

model, this function is affine, so spot interest rates are perfectly (linearly) correlated. Empirical observation suggests that spot rates of different maturities are positively but not perfectly correlated.[234] The perfect correlation between spot interest rates is one of the most serious drawbacks of one-factor (affine) interest rate models.[235]

Zero-coupon bond prices are non-linear functions of spot rates.[236] Their correlation structure is therefore quite similar to the spot rate correlation structure. In the Vasicek model, they are perfectly – but non-linearly – correlated. The correlation matrix Ψ of the risky zero-coupon bonds is in our numerical example[237]

$$\Psi = \begin{pmatrix} 1 & 0.99 & 0.99 & 0.99 & 0.99 & 0.99 & 0.99 & 0.99 & 0.99 \\ 0.99 & 1 & 0.99 & 0.99 & 0.99 & 0.99 & 0.99 & 0.99 & 0.99 \\ 0.99 & 0.99 & 1 & 0.99 & 0.99 & 0.99 & 0.99 & 0.99 & 0.99 \\ 0.99 & 0.99 & 0.99 & 1 & 0.99 & 0.99 & 0.99 & 0.99 & 0.99 \\ 0.99 & 0.99 & 0.99 & 0.99 & 1 & 0.99 & 0.99 & 0.99 & 0.99 \\ 0.99 & 0.99 & 0.99 & 0.99 & 0.99 & 1 & 0.99 & 0.99 & 0.99 \\ 0.99 & 0.99 & 0.99 & 0.99 & 0.99 & 0.99 & 1 & 0.99 & 0.99 \\ 0.99 & 0.99 & 0.99 & 0.99 & 0.99 & 0.99 & 0.99 & 1 & 0.99 \\ 0.99 & 0.99 & 0.99 & 0.99 & 0.99 & 0.99 & 0.99 & 0.99 & 1 \end{pmatrix} \tag{4.20}$$

Future zero-coupon bond prices are lognormally distributed because the short rate is normally distributed.[238] The expected values and standard deviations of zero-coupon bond prices of different maturities $(1, \ldots, 10)$ at time $T = 1$ are given in the next table.

Table 4.3. Vasicek model: Distribution of zero-coupon bond prices.

T_i	1	2	3	4	5	6	7	8	9	10
$E_0[P(1,T_i)]$	1.000	0.974	0.946	0.917	0.888	0.858	0.829	0.801	0.772	0.745
$\text{std}_0(P(1,T_i))$	0.000	0.013	0.023	0.031	0.037	0.041	0.044	0.047	0.048	0.049

With current prices $P(0, T_i)$ we can calculate the (continuously compounded) expected returns of the zero-coupon bonds over the next period.

234 See Golub/Tilman (2000), p. 89.
235 See Martellini/Priaulet/Priaulet (2003), p. 391.
236 See Equation (3.28).
237 The correlation matrix can be determined from the covariance matrix that was defined in Equation (4.6). We truncated the correlations at two decimal places for presentational purposes.
238 See Equation (3.28).

Table 4.4. Vasicek model: Expected holding period returns.

T_i	1	2	3	4	5	6	7	8	9	10
Exp. ret. (%)	2.716	2.975	3.18	3.345	3.477	3.584	3.671	3.743	3.802	3.85

In order to gain some intuition about possible portfolio returns, we perform a simple scenario analysis. The different scenarios are determined by the value of the short rate at time 1 only.[239] When the short rate rises (falls), longer-term bonds perform worse (better) than short-term bonds. When the term structure stays the same (i.e. r(1)=r(0)=0.025) then every zero-coupon bond with the maturity T_i earns the forward rate between $T_i - 1$ and T_i.[240]

Table 4.5. Vasicek model: Zero-coupon bond returns in different scenarios.

T_i	1	2	3	4	5	6	7	8	9	10
$r(1) = 0.030$	2.716	2.562	2.419	2.288	2.17	2.066	1.974	1.894	1.824	1.764
$r(1) = 0.025$	2.716	3.023	3.269	3.468	3.63	3.762	3.87	3.959	4.033	4.094
$r(1) = 0.020$	2.716	3.483	4.119	4.648	5.089	5.457	5.766	6.024	6.241	6.423
$r(1) = 0.015$	2.716	3.944	4.97	5.829	6.549	7.153	7.661	8.089	8.449	8.753
$r(1) = 0.010$	2.716	4.404	5.82	7.009	8.008	8.849	9.557	10.15	10.66	11.08
$r(1) = 0.005$	2.716	4.865	6.67	8.189	9.467	10.54	11.45	12.22	12.87	13.41

These tables were provided for assessing whether or not the term structure parameter values in Table 4.2 are sensible. Huge percentage gains of individual zero-coupon bonds would have attracted attention in Table 4.5. We conclude that the parameters provide for realistic zero-coupon bond returns.

Next, we turn our attention to mean-variance efficient portfolios and obtain the tangency portfolio. For an initial wealth of $W_0 = 1$ unit of account, the tangency portfolio $N_{\tan}$ is[241]

$$N_{\tan} = \begin{pmatrix} 19.06 \\ -155.91 \\ 735.31 \\ -2198.36 \\ 4312.56 \\ -5543.49 \\ 4497.95 \\ -2088.91 \\ 422.85 \end{pmatrix} \tag{4.21}$$

[239] Of course the probability for the short rate to be exactly x is equal to zero, but nevertheless we give here 5 scenarios for 5 different values of the short rate.

[240] See e.g. Ilmanen (1995), p. 4.

[241] The tangency portfolio is defined in Equation (4.13).

It contains enormous long and short positions primarily in the 6-, 7- and 8-year zero-coupon bonds.[242] According to our solution the investor is supposed to buy 4313 units of account of face value of the 6-year zero-coupon bond for every unit of account of initial wealth. The reason for the unrealistic portfolio composition is the following: A profound inspection of the tangency portfolio reveals that the sign of the position alternates, i.e. a long position in the T_i-year zero-coupon bond is followed by a short position in the $T_i + 1$-year zero-coupon bond (and vice versa). Furthermore, for adjacent maturities the correlations are highest. These zero-coupon bonds are therefore considered near perfect substitutes (from a diversification perspective) and so (because of differences in expected values and standard deviations) the optimization procedure tries to exploit a suspected arbitrage opportunity by buying one zero-coupon bond and going short the other.

In order to obtain portfolios that can be implemented in practice, we prohibit short sales.[243] A drawback of this measure is the lack of an analytic solution for the bond portfolio optimization problem. Another option would be to restrict the number of zero-bonds the investor can buy. This is suggested by Korn/Koziol (2006). They propose to consider only so many bonds as the term structure model has risk factors.[244] They give however no guidance on which bond(s) to select, so this methods seems ad-hoc and is not considered here.

Table 4.6 contains the portfolio weights for the ten zero-coupon bonds for different expected portfolio values.[245]

---

[242] The analysis by Korn/Koziol (2006) finds similar portfolios.

[243] This is a common investment constraint for mutual funds in practice. § 59 InvG (German Investment Act) forbids for example short sales for German mutual funds.

[244] See Korn/Koziol (2006), p. 22.

[245] The gray column indicates the position in the riskless zero-coupon bond. The constrained optimization problem was solved with Wolfram Research's Mathematica 5.2 package using the NMinimize function.

Table 4.6. Vasicek model: Zero-coupon bond weights for short-sale constrained portfolios.

$E[W_1]$	w_1	w_2	w_3	w_4	w_5	w_6	w_7	w_8	w_9	w_{10}
1.028	1	0	0	0	0	0	0	0	0	0
1.029	0.44	0.56	0	0	0	0	0	0	0	0
1.031	0	0.85	0.15	0	0	0	0	0	0	0
1.032	0	0.18	0.82	0	0	0	0	0	0	0
1.034	0	0	0.45	0.55	0	0	0	0	0	0
1.035	0	0	0	0.58	0.42	0	0	0	0	0
1.037	0	0	0	0	0.59	0.41	0	0	0	0
1.038	0	0	0	0	0	0.41	0.59	0	0	0
1.040	0	0	0	0	0	0	0.37	0.14	0.49	0
1.042	0	0	0	0	0	0	0	0	0	1

These mean-variance efficient portfolios contain at most three zero-coupon bonds. This is not surprising since all assets are nearly perfectly correlated. Any given expected value can therefore be obtained by a linear combination of just two assets since adding more assets wouldn't diversify the portfolio much more.[246] Nevertheless, it is interesting to note, that the portfolios consist of different bonds and not only positions in the long (maximum maturity) and short (minimum maturity) bond for example.

Table 4.7 compares the standard deviations of terminal wealth for the unconstrained and the short-sale constrained case. Although the portfolio composition differs significantly, the difference in the portfolio standard deviation for the same expected terminal wealth is small.

Table 4.7. Vasicek model: Standard deviations of terminal wealth for unconstrained and short-sale constrained portfolios

$E[W_1]$	unconstrained	constrained
1.028	0.0000	0.0000
1.029	0.0075	0.0076
1.031	0.0149	0.0151
1.032	0.0224	0.0227
1.034	0.0299	0.0303
1.035	0.0374	0.0379
1.037	0.0449	0.0456
1.038	0.0523	0.0532
1.040	0.0598	0.0609
1.042	0.0673	0.0685

[246] Recall the two asset Markowitz diagram for varying correlations in classic finance textbooks, e.g. Elton et al. (2003), p. 77

As can be seen from Table 4.7, the loss in "efficiency" (i.e. higher standard deviation) due to the introduction of short sale constraints is negligible. This is not surprising since we already mentioned that the assets are nearly perfectly correlated and are therefore quite perfect substitutes. Constraining positions in these assets forces switching but this forced switching is relatively costless for the investor, i.e. it raises the portfolio standard deviation only slightly.

Next, we analyze the long-term portfolio selection problem.

Long-term investment horizon

If the investment horizon is greater than one period, then the portfolio value at the investment horizon is not only determined by the term structure at time T but also by the term structures at every date before T, because all cash flows that are received before the investment horizon are reinvested at the current spot rate until T. Changes in the short rate have therefore radically different effects on short[247] and long bonds[248] – analogous to the classical duration analysis (reinvestment risk versus market risk).[249] The major change from the short-term to long-term case therefore is the correlation matrix Ψ of the risky zero-coupon bonds[250]

$$\Psi = \begin{pmatrix} 1 & 0.76 & 0.67 & 0.62 & -0.59 & -0.59 & -0.59 & -0.59 & -0.59 \\ 0.76 & 1 & 0.88 & 0.81 & -0.77 & -0.77 & -0.77 & -0.77 & -0.77 \\ 0.67 & 0.88 & 1 & 0.93 & -0.88 & -0.88 & -0.88 & -0.88 & -0.88 \\ 0.62 & 0.81 & 0.93 & 1 & -0.95 & -0.95 & -0.95 & -0.95 & -0.95 \\ -0.59 & -0.77 & -0.88 & -0.95 & 1 & 0.99 & 0.99 & 0.99 & 0.99 \\ -0.59 & -0.77 & -0.88 & -0.95 & 0.99 & 1 & 0.99 & 0.99 & 0.99 \\ -0.59 & -0.77 & -0.88 & -0.95 & 0.99 & 0.99 & 1 & 0.99 & 0.99 \\ -0.59 & -0.77 & -0.88 & -0.95 & 0.99 & 0.99 & 0.99 & 1 & 0.99 \\ -0.59 & -0.77 & -0.88 & -0.95 & 0.99 & 0.99 & 0.99 & 0.99 & 1 \end{pmatrix}$$

Next, we calculate the tangency portfolio as in the last section. For an initial wealth of $W_0 = 1$ units of account, the tangency portfolio N_{tan} is

$$N_{\text{tan}} = \begin{pmatrix} 0 \\ 0 \\ 0 \\ 0 \\ 8.14 \\ -22.70 \\ 31.51 \\ -21.74 \\ 5.96 \end{pmatrix} \tag{4.22}$$

---

[247] Maturity less than T.

[248] Maturity greater than T.

[249] See Garbade (1996), pp. 37–40.

[250] The correlation matrix is truncated to two decimal places for presentational purposes.

In comparison to the short-term case[251], this tangency portfolio still contains substantial long and short positions. For longer maturity bonds, the alternating signs that we discovered in last section's tangency portfolio are also visible. This is not surprising since as can be seen from the above correlation matrix, long term bonds are highly correlated among each other. We pick up the argumentation from the short-term case and restrict short sales in order to obtain portfolios that could be implemented in practice.

The mean-variance efficient portfolios with short-sale constraints are shown in Table 4.8.[252]

Table 4.8. Vasicek model: Zero-coupon bond weights for short-sale constrained portfolios.

$E[W_5]$	w_1	w_2	w_3	w_4	w_5	w_6	w_7	w_8	w_9	w_{10}
1.168	0	0	0	0	1	0	0	0	0	0
1.172	0.02	0.03	0.06	0.14	0.08	0.66	0	0	0	0
1.176	0.04	0.07	0.14	0.09	0	0.03	0.63	0	0	0
1.181	0.06	0.11	0.19	0	0	0	0	0.64	0	0
1.185	0.07	0.17	0.12	0	0	0	0	0	0.64	0
1.189	0.08	0.22	0.04	0	0	0	0	0	0	0.65
1.194	0.08	0.19	0	0	0	0	0	0	0	0.74
1.198	0.06	0.11	0	0	0	0	0	0	0	0.83
1.202	0.05	0.04	0	0	0	0	0	0	0	0.91
1.207	0	0	0	0	0	0	0	0	0	1

In contrast to the short-term case considered in the last section, the investor holds more zero-coupon bonds (up to six). The structure of the bond portfolios differ also. In the short-term case, the maturity of the zero-coupon bonds was concentrated at one point on the maturity spectrum.[253] The long-term case produces efficient portfolio that consist of short and long bonds but no positions in intermediate bonds.

Next, we compare the short-sale constrained optimization to the unconstrained optimization in terms of differences in portfolio standard deviation for specific expected values of terminal wealth.

---

[251] See Equation (4.21).

[252] The gray column indicates the position in the riskless zero-coupon bond. The constrained optimization problem was solved with Wolfram Research's Mathematica 5.2 package using the NMinimize function.

[253] For a given expected value of terminal wealth.

Table 4.9. Vasicek model: Standard deviations of terminal wealth for unconstrained and short-sale constrained portfolios

$E[W_5]$	unconstrained	constrained
1.16775	0.0000	0.0000
1.17208	0.0095	0.0101
1.17641	0.0193	0.0204
1.18075	0.0290	0.0309
1.18508	0.0386	0.0416
1.18941	0.0483	0.0524
1.19374	0.0579	0.0634
1.19808	0.0676	0.0748
1.20241	0.0773	0.0862
1.20674	0.0869	0.0978

The result of the comparison is quite the same as in the short-term investment horizon case. The rise in portfolio standard deviation due to short-sale constraints is negligible, although the portfolio compositions vary significantly.

In this section we have shown how the mean-variance approach can be used for bond portfolio selection. The bond market was governed by the Vasicek term structure model. An unconstrained optimization yielded portfolios with huge long and short positions. Since these can't be implemented in the real world, we constrained short sales. The resulting portfolios were much more plausible. Another interesting point is that the introduction of short sale constraints didn't have a significant influence on the portfolio standard deviations. In the next section we analyze the results of the HW2 model.

4.2.5 Two-Factor Hull/White (1994) Model

We derived the HW2 model in Chapter 3.5. It assumes two bivariate normally distributed state variables, $r(t)$ and $\varepsilon(t)$. The bond portfolio selection parameters can then be calculated easily since this model is a special case of the multi-factor Gaussian setting outlined in Chapter 4.2.2 with $x(t)' = (r(t), \varepsilon(t))$ and

$$B(t,T)' = (B_1(t,T), B_2(t,T))$$

The covariance matrices are then defined as follows:

$$\mathrm{COV}(x(t), x(t)') = \begin{pmatrix} \mathrm{var}(r(t)) & \mathrm{cov}(r(t), \varepsilon(t)) \\ \mathrm{cov}(\varepsilon(t), r(t)) & \mathrm{var}(\varepsilon(t)) \end{pmatrix}$$

and

$$\mathrm{COV}(x(t), x(\tau)') = \begin{pmatrix} \mathrm{cov}(r(t), r(\tau)) & \mathrm{cov}(r(t), \varepsilon(\tau)) \\ \mathrm{cov}(\varepsilon(t), r(\tau)) & \mathrm{cov}(\varepsilon(t), \varepsilon(\tau)) \end{pmatrix}$$

Hull/White (1996) argue that, when considering a two-factor model, using a simple best-fit calibration technique to match the current term structure of interest rates is not the right choice, instead one should set some parameters to values that make economic sense and only fit the remaining parameters to the current term structure of interest rates.[254]

We adopt this approach and assume economically plausible values for the current spot rate $r(0)$ and for the correlation coefficient ϱ.[255] Table 4.10 summarizes the parameter values

Table 4.10. HW2 model: Parameter values for numerical example.

Parameter	Value
r(0)	0.025
$\varepsilon(0)$	0
ϱ	0.6
θ	0.0053
κ_r	0.2591
κ_ε	0.8274
σ_r	0.0073
σ_ε	0.0219
λ_1	1.2395
λ_2	0

This choice of parameters approximates the German term structure of interest rates as of January 2006 quite good and furthermore results in a plausible correlation structure.[256]

Short-term investment horizon

First, we want to examine the case $T = 1$. The major difference between one-factor and two-factor models is the correlation structure between spot interest rates of different maturity.[257]

Given the above parameters, the HW2 model yields the following correlation matrix for the zero-coupon bonds.[258]

[254] See Hull/White (1996), p. 289.

[255] The HW2 model assumes $\varepsilon(0) = 0$.

[256] As shown in Equation (4.23).

[257] In a one factor affine model, all correlations are equal to 1.

[258] As usual it is truncated for presentational purposes to two decimal places.

$$\Psi = \begin{pmatrix} 1 & 0.9 & 0.72 & 0.58 & 0.48 & 0.42 & 0.38 & 0.35 & 0.33 \\ 0.9 & 1 & 0.95 & 0.87 & 0.81 & 0.77 & 0.74 & 0.72 & 0.71 \\ 0.72 & 0.95 & 1 & 0.98 & 0.96 & 0.93 & 0.91 & 0.9 & 0.89 \\ 0.58 & 0.87 & 0.98 & 1 & 0.99 & 0.98 & 0.97 & 0.97 & 0.96 \\ 0.48 & 0.81 & 0.96 & 0.99 & 1 & 0.99 & 0.99 & 0.99 & 0.99 \\ 0.42 & 0.77 & 0.93 & 0.98 & 0.99 & 1 & 0.99 & 0.99 & 0.99 \\ 0.38 & 0.74 & 0.91 & 0.97 & 0.99 & 0.99 & 1 & 0.99 & 0.99 \\ 0.35 & 0.72 & 0.9 & 0.97 & 0.99 & 0.99 & 0.99 & 1 & 0.99 \\ 0.33 & 0.71 & 0.89 & 0.96 & 0.99 & 0.99 & 0.99 & 0.99 & 1 \end{pmatrix} \tag{4.23}$$

The correlation matrix a two-factor model captures the real-world correlations of the bond market quite well. Zero-coupon bond prices of different maturities are positively but not perfectly correlated. The higher the maturity difference between zero-coupon bonds, the lower is the correlation between them, e.g. $\mathrm{corr}(P(1,2), P(1,3)) = 0.9$ and $\mathrm{corr}(P(1,2), P(1,10)) = 0.33$. But for a fixed maturity difference (e.g. 1 period) the correlations get higher the more the maturity differs from the investment horizon, e.g. $\mathrm{corr}(P(1,2), P(1,3)) = 0.9$ but $\mathrm{corr}(P(1,6), P(1,7)) \approx 1$.

Figures 4.3 and 4.4 show the current (and expected) term structures as well as the term structure of volatility.

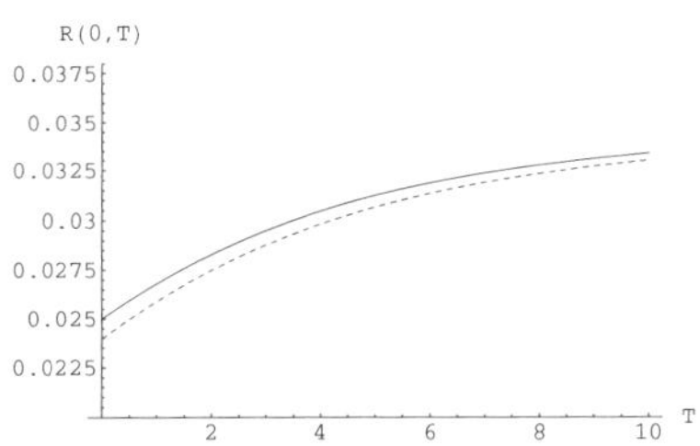

Fig. 4.3. HW2: Term structure

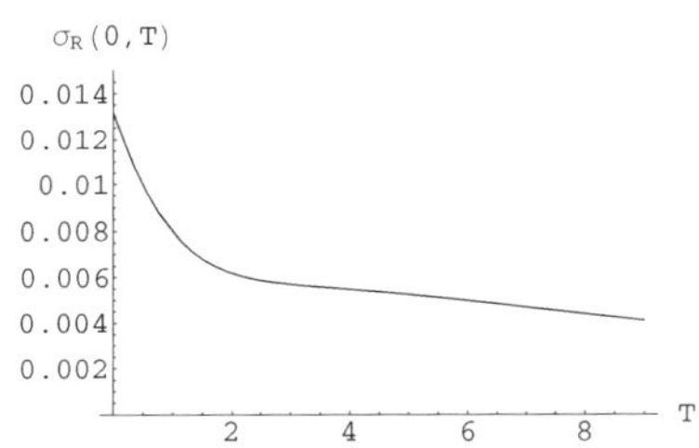

Fig. 4.4. HW2: Volatility structure

First we derive the expected returns of zero-coupon bonds of different maturities. These are shown in Table 4.11. We observe that an investment in longer term zero-coupon bonds promises a higher expected return. This is not surprising since interest rates are expected to decline[259] and therefore longer term bonds are expected to gain more.

Table 4.11. HW2 model: Expected holding period returns of zero-coupon bonds.

T_i	1	2	3	4	5	6	7	8	9	10
Exp. ret. (%)	2.681	3.071	3.354	3.553	3.69	3.785	3.852	3.899	3.934	3.958

[259] More precisely a nearly parallel downward movement for the relevant maturities. See Figure 4.3.

In order to derive mean-variance efficient portfolios, we calculate the tangency portfolio. It contains – as has been the case in the Vasicek model – very large long and short positions.

$$N_{\text{tan}} = \begin{pmatrix} 28.98 \\ -325.55 \\ 1840.40 \\ -5314.76 \\ 5932.30 \\ 6187.92 \\ -24694.20 \\ 25114.00 \\ -8781.28 \end{pmatrix} \tag{4.24}$$

Since constructing such a portfolio is highly unrealistic in practice, we concentrate once again on the short-sale restricted case. We give the portfolio weights for different expected values of terminal wealth in Table 4.12.[260]

Table 4.12. HW2 model: Zero-coupon bond weights for short-sale constrained portfolios.

$E[W_1]$	w_1	w_2	w_3	w_4	w_5	w_6	w_7	w_8	w_9	w_{10}
1.027	1	0	0	0	0	0	0	0	0	0
1.029	0.76	0.08	0.12	0.04	0	0	0	0	0	0
1.030	0.56	0	0.44	0	0	0	0	0	0	0
1.032	0.25	0.36	0.23	0.18	0	0	0	0	0	0
1.033	0.12	0	0.88	0	0	0	0	0	0	0
1.035	0	0	0.67	0.33	0	0	0	0	0	0
1.036	0	0	0	0.91	0.09	0	0	0	0	0
1.038	0	0	0	0	0.86	0.14	0	0	0	0
1.040	0	0	0	0	0	0.24	0.76	0	0	0
1.041	0	0	0	0	0	0	0	0	0	1

For lower expected values of terminal wealth, the portfolios are quite diversified with up to four zero-coupon bonds contained in the portfolio. For higher expected values, the portfolios contain at most two zero-coupon bonds. Diversification is apparently only possible for lower expected values. A further look at the correlation matrix clarifies this result. Long-term bonds are highly correlated among each other and these are the bonds that offer the higher expected returns necessary for high expected terminal wealth. In order to achieve high expected returns, the investor has to buy long-term bonds but then the diversification potential vanishes because these bonds are highly correlated

260 The gray column indicates the riskless zero-coupon bond. The constrained optimization problem was solved with Wolfram Research's Mathematica 5.2 package using the NMinimize function.

and therefore only two bonds are included in the optimal portfolio. For lower expected portfolio returns, short and long bonds can be bought and between these "groups", the correlation is quite low. The structure of the portfolios resembles the short-term Vasicek case, because normally no position is taken in extreme maturity sectors.

Next, we again examine the loss in diversification due to short-sale constraints. Table 4.13 contains the portfolio standard deviations for the unconstrained and the short-sale constrained optimal portfolios for different expected values.

Table 4.13. HW2 model: Standard deviations of terminal wealth for unconstrained and short-sale constrained portfolios

$E[W_1]$	unconstrained	constrained
1.027	0.0000	0.0000
1.029	0.0005	0.0028
1.030	0.0010	0.0056
1.032	0.0016	0.0085
1.033	0.0021	0.0113
1.035	0.0026	0.0143
1.036	0.0031	0.0182
1.038	0.0037	0.0234
1.040	0.0042	0.0302
1.041	0.0047	0.0386

It can be seen from this table that in the HW2 model, when considering a short-term investment horizon, there are sizeable increases in portfolio risk due to the introduction of short-sale constraints.

Long-term investment horizon

For a long-term investment horizon ($T = 5$) the results turn out to be nearly identical to the short term case although some correlations are now negative.[261] The unconstrained portfolios contain again large long and short positions, so we will give the results for the short-sale constrained optimization only. Table 4.14 gives the portfolio weights for the assets in the short-sale constrained case.[262]

[261] This has also been the case in the long-term Vasicek model.

[262] The gray column indicates the riskless zero-coupon bond. The constrained optimization problem was solved with Wolfram Research's Mathematica 5.2 package using the NMinimize function.

Table 4.14. HW2 model: Zero-coupon bond weights for short-sale constrained portfolios.

$E[W_5]$	w_1	w_2	w_3	w_4	w_5	w_6	w_7	w_8	w_9	w_{10}
1.169	0	0	0	0	1	0	0	0	0	0
1.174	0	0	0	0	0.85	0	0.01	0.14	0	0
1.178	0	0	0	0	0.71	0	0.02	0.27	0	0
1.183	0	0	0	0	0.56	0	0.03	0.41	0	0
1.187	0	0	0	0	0.41	0	0.04	0.55	0	0
1.192	0	0	0	0	0.26	0	0.05	0.69	0	0
1.196	0	0	0	0	0.11	0	0.07	0.82	0	0
1.200	0	0	0	0	0	0	0	0.92	0.08	0
1.205	0	0	0	0	0	0	0	0.09	0.91	0
1.209	0	0	0	0	0	0	0	0	0	1

It is interesting to note that no positions are held in short bonds, i.e. no position is taken in bonds with a maturity shorter than 5 years. In this example – from a mean-variance perspective – rolling over, i.e. investing in short bonds and reinvesting until the investment horizon, seems to be an inefficient strategy. Furthermore, it is conspicuous that quite large positions are held in the riskless bond.[263] Regarding the structure of the portfolios, the outcome resembles the short-term case, i.e. the resulting portfolios resemble bullet portfolios.

Table 4.15 summarizes the mean-variance information of the constrained and unconstrained portfolios.

Table 4.15. HW2 model: Standard deviations of terminal wealth for unconstrained and short-sale constrained portfolios

$E[W_5]$	unconstrained	constrained
1.169	0.0000	0.0000
1.174	0.0032	0.0105
1.178	0.0065	0.0210
1.183	0.0097	0.0314
1.187	0.0130	0.0419
1.192	0.0162	0.0524
1.196	0.0194	0.0629
1.200	0.0227	0.0734
1.205	0.0259	0.0844
1.209	0.0292	0.0961

As can be seen from Table 4.15, there are sizeable increases in portfolio risk due to the introduction of short-sale constraints.

[263] This is in contrast to the findings for the Vasicek model. There, nearly no positions were taken in the riskless bond, see Table 4.8.

In this section the bond market was governed by the HW2 term structure model. An unconstrained optimization yielded portfolios with large long and short positions. Hence, we again constrained short sales. The resulting portfolios were much more plausible. In contrast to the findings in the Vasicek model, the introduction of short sale constraints had a significant effect on the portfolio standard deviations.

In the next part, we want to analyze how mean-variance efficient portfolios compare to portfolios resulting from active and passive bond portfolio strategies employed in practice in various interest rate scenarios.

4.3 Static Bond Portfolio Selection in Practice

4.3.1 Introduction

Despite its broad acceptance in the equity markets (by both academics and practitioners alike), the mean-variance framework could never quite establish a foothold in the fixed-income markets.[264] This might be due to the added complexity as has become apparent from the analysis in the last section or as Wilhelm (1992) suggests, due to low interest rate volatility in the 1970s.[265]

In this chapter we want to compare common bond portfolio selection techniques used in practice with mean-variance efficient portfolios resulting from the framework introduced in the last section by means of numerical examples.[266] Since there exist numerous bond portfolio selection methods in practice, we have to select methods that are representative and lend itself easily to a comparison with the mean-variance approach.

Bond portfolio selection strategies can be classified in various ways. A common approach is to distinguish between active and passive portfolio selection strategies.[267] An active strategy can be defined as an investment strategy that requires substantial expectational input, i.e. in our context, the investor must have a "view" on the term structure evolution.[268] Passive strategies on the other hand require no or only minimal expectational input.[269] In the course of this section we will examine both strategies. We analyze active bond portfolio selection strategies first.

264 The classic book by Fabozzi (2004) doesn't even mention mean-variance analysis as a possible portfolio selection tool.

265 See Wilhelm (1992), p. 210.

266 For a complete treatise on bond portfolio selection see e.g. Fabozzi (2000), Fabozzi (2001), Fabozzi (2004) and Martellini/Priaulet/Priaulet (2003).

267 See Martellini/Priaulet/Priaulet (2003), p. 211.

268 See Fabozzi (2004), p. 412.

269 See Fabozzi (2004), p. 412.

4.3.2 Active Bond Portfolio Selection Strategies

The practice generally distinguishes at least three active bond portfolio selection strategies:[270]

- Riding/rollover strategies
- Duration strategies and
- Yield curve strategies

Riding the Yield Curve and Rollover Strategies

Riding the yield curve entails buying longer dated bonds[271] and selling them before maturity.[272] Rollover refers to the strategy of buying shorter dated bonds, holding them until maturity and investing the proceeds again.[273] Both strategies rely on the view that the current term structure of interest rates remains relatively stable.[274]

According to the practice, the selection of the strategy (riding vs. rollover) depends on the slope of the yield curve.[275] When the term structure does not change[276], then every zero-coupon bond with the maturity T_i has a Δt-holding period return equal to the forward rate between $T_i - \Delta t$ and T_i.[277] The forward rates are at the same time so-called break-even rates. They show how much rates can increase before the investor earns less than the riskless spot rate.[278]

In a positive term structure environment, investors hence utilize the riding strategy, because then the holding period return of the longer dated bonds is greater simply because the maturity shortens.[279] On the other hand, if the term structure is downward sloping, then the investor can gain higher returns if he employs the rollover strategy. Since the yield curve is upward sloping most of the time, we concentrate on the riding strategy.

The empirical results on the profitability of the riding strategy are mixed.[280] A recent study by Bieri/Chincarini (2005) examines which signal – e.g. positive slope of yield curve – should be used to initiate a riding strategy. They find that investors historically could have significantly enhanced their returns by riding the yield curve instead of buying and holding zero-coupon bonds with a maturity equal to their investment horizon.[281] From a mean-variance

[270] See Martellini/Priaulet/Priaulet (2003), pp. 234–240.
[271] Maturity longer than the investment horizon.
[272] See Bieri/Chincarini (2005), p. 6.
[273] See Martellini/Priaulet/Priaulet (2003), p. 237.
[274] See Martellini/Priaulet/Priaulet (2003), p. 234.
[275] See Martellini/Priaulet/Priaulet (2003), p. 234.
[276] $R(t,T) = R(t + \Delta T, T + \Delta t) \; \forall \; T$
[277] See Ilmanen (1995), p. 4.
[278] See Bieri/Chincarini (2005), p. 9.
[279] See Zimmerer (2003), p. 243.
[280] For a summary of their findings see Martellini/Priaulet/Priaulet (2003), p. 235.
[281] See Bieri/Chincarini (2005), p. 28.

perspective one could argue that the deviation of the asset maturities from the investment horizon in the riding strategy makes the strategy riskier, so one can expect to earn a higher return from taking on the risk. Furthermore, the strategy doesn't give advice on the bond[282] with which to implement the strategy. A naive approach would be to buy the zero-coupon bond with the biggest expected holding period return. It becomes obvious that the concept of diversification is not part of the strategy. If the investor thinks that riding is preferable to buying and holding, he invests all his money in the respective bond.

We now compare – by means of a numerical example – the riding strategy with the mean-variance framework. In the Vasicek model, the expected future term structure depends only on the short-rate, so when the expected future short-rate is the same as today, the whole term-structure is expected to stay the same. As we pointed out in Chapter 3.4 this is the case when $r(0) = \theta$, i.e. the current short rate is equal to the mean-reversion level. We assume an investment horizon of $T = 1$ and an investment universe consisting of $\tau = 10$ zero-coupon bonds of different maturities $1, \ldots, \tau$. Furthermore, we assume the following parameters for the Vasicek term structure model

Table 4.16. Vasicek model: Riding strategy parameter values.

Parameter	Value
r(0)	0.024
θ	0.024
κ	0.1668
σ_r	0.0153
λ	0.2126

The term structure of interest rates is then positively sloped[283] and expected to remain unchanged, because $r(t) = \theta$. Table 4.17 contains the portfolio weights for mean-variance efficient bond portfolios for different expected values of terminal wealth.

---

[282] It is always a single bond that is bought.

[283] Because $r(t) \leq R(\infty) - \frac{\sigma_r^2}{4\kappa^2}$, see Vasicek (1977), p. 168.

Table 4.17. Vasicek model: Portfolio weights when term structure is expected to remain unchanged.

$E[W_1]$	w_1	w_2	w_3	w_4	w_5	w_6	w_7	w_8	w_9	w_{10}
1.026	1	0	0	0	0	0	0	0	0	0
1.027	0.44	0.56	0	0	0	0	0	0	0	0
1.029	0	0.85	0.15	0	0	0	0	0	0	0
1.030	0	0.18	0.82	0	0	0	0	0	0	0
1.032	0	0	0.43	0.57	0	0	0	0	0	0
1.034	0	0	0	0.58	0.42	0	0	0	0	0
1.035	0	0	0	0.1	0.38	0.52	0	0	0	0
1.037	0	0	0	0	0	0.53	0.33	0.14	0	0
1.038	0	0	0	0	0	0	0	0.95	0.05	0
1.040	0	0	0	0	0	0	0	0	0	1

The highlighted row in Table 4.17 refers to the buy-and-hold strategy. All other rows contain the portfolio weights of short-sale restricted mean-variance efficient portfolios. It is interesting that the optimum portfolios are close to the ones proposed by the riding strategy.[284] The portfolio weights are concentrated at specific points on the maturity spectrum. But the concentration is not perfect, i.e. the investor should not buy one single zero-coupon bond but at least two.[285] But – as has been noted before – this being a one-factor model, the diversification potential is quite limited. So, in practice the difference in terms of portfolio volatility between buying a single zero-coupon bond or a portfolio of bonds with the same expected value is negligible.[286] Since the results of this analysis are very similar to the numerical example in the last section, we do not analyze the HW2 model.

The mean-variance framework generates portfolios that are comparable to those recommended by the riding strategy. It offers furthermore the advantage that the investor can assess the riskiness of the different possible implementations (zero-coupon bonds to buy) of the strategy better.

Future research could analyze how the riding strategy performs in comparison to mean-variance efficient portfolios in different scenarios (also when the expectations are not met). We suspect that mean-variance efficient portfolios would turn out to be superior, because in this framework the uncertainty regarding the outcome is already being accounted for.

Duration strategies

Naive duration strategies require only expectations about one variable, the level of interest rates. They assume, that only one factor drives the term

[284] Longer dated bonds are bought.

[285] This result is not due to the specific expected values chosen, another run with 50 different expected portfolio values resulted in similarly diversified portfolios.

[286] See Chapter 4.2.4.

structure and this factor affects all rates in the same way, i.e. the term structure is affected only by parallel movements.[287]

It is well known that – inside the duration framework – the investor can implement his view about level changes by setting the portfolio's Macaulay duration relative to his investment horizon. If he expects interest rates to fall (increase), he chooses a portfolio with a duration longer (shorter) than the investment horizon. A comparison of the duration strategies and the mean-variance framework is not perfect since – at least in the Vasicek model – there can be no parallel movements of the term structure of interest rates.[288] Longer-term rates change less than shorter-term interest rates.[289] We now want to present two numerical examples and compare the outcomes to the advice given by the duration strategy. We have to focus on the long term case (investment horizon is 5 years) since in the short term case portfolio durations smaller than the investment horizon are impossible when short-sales are not allowed.

Falling Interest Rates. In Chapters 4.2.4 and 4.2.5 we presented a numerical example for the Vasicek and the HW2 model, where the interest rates were expected to fall.[290] Hence, the duration strategy would advise to construct a portfolio with a duration greater than the investment horizon, i.e. greater than 5. A closer look at Tables 4.8 and 4.14 reveals that the portfolio durations are in all cases[291] greater than the investment horizon. The result of the (short-sale constrained) mean-variance optimization is hence in line with the practical duration advice.

Rising Interest Rates. Next we consider the case where the investor expects interest rates to rise. We adapt the parameters from Table 4.16 in order to obtain an expected rise in the term structure of interest rates. We assume the following parameters for the Vasicek model

Table 4.18. Vasicek model: Duration strategy parameter values.

Parameter	Value
r(0)	0.02
θ	0.024
κ	0.1668
σ_r	0.0153
λ	0.2126

287 See Martellini/Priaulet/Priaulet (2003), p. 236.

288 See Equation (3.32) and Table 3.1. In the Ho/Lee (1986) model however, only parallel term structure movements can occur.

289 In the Vasicek model the infinitely long rate stays constant.

290 See Figures 4.1 and 4.3.

291 Disregarding the riskless investment option.

The current term structure of interest rates is hence upward-sloping (solid line) and is expected to rise (dashed line) as can be seen from Figure 4.5

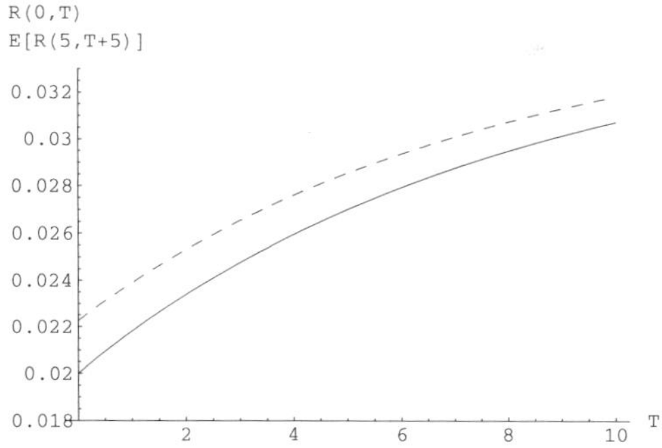

Fig. 4.5. Vasicek model: Expected term structure

The duration strategy would advise to construct a portfolio with a duration smaller than the investment horizon. The portfolio optimization results are given in Table 4.19.

Table 4.19. Vasicek model: Portfolio weights and durations for duration strategy comparison.

$E[W_5]$	w_1	w_2	w_3	w_4	w_5	w_6	w_7	w_8	w_9	w_{10}	MacDur.
1.145	0	0	0	0	1	0	0	0	0	0	5.00
1.149	0.02	0.03	0.06	0.14	0.08	0.66	0	0	0	0	5.21
1.153	0.04	0.07	0.15	0.04	0	0.17	0.53	0	0	0	5.52
1.157	0.06	0.11	0.19	0	0	0	0	0.64	0	0	5.95
1.162	0.07	0.17	0.12	0	0	0	0	0	0.64	0	6.53
1.166	0.08	0.22	0.04	0	0	0	0	0	0	0.65	7.18
1.170	0.08	0.18	0	0	0	0	0	0	0	0.74	7.84
1.174	0.06	0.11	0	0	0	0	0	0	0	0.83	8.54
1.179	0.05	0.04	0	0	0	0	0	0	0	0.91	9.24
1.183	0	0	0	0	0	0	0	0	0	1	10.00

Surprisingly the portfolio (Macaulay) duration is in all relevant cases greater than the investment horizon, therefore contradicting the advice given by duration strategy. This result can be reproduced in a similar term structure environment with the HW2 model. We assume the following parameters:

Table 4.20. HW2 model: Duration strategy parameter values.

Parameter	Value
r(0)	0.018
$\varepsilon(0)$	0
ϱ	0.6
θ	0.0053
κ_r	0.2591
κ_ε	0.8274
σ_r	0.0073
σ_ε	0.0219
λ_1	1.2395
λ_2	0

Table 4.21 summarizes the portfolio weights and the Macaulay portfolio durations for different expected values of terminal wealth

Table 4.21. HW2 model: Portfolio weights and Macaulay durations for duration strategy comparison.

$E[W_5]$	w_1	w_2	w_3	w_4	w_5	w_6	w_7	w_8	w_9	w_{10}	MacDur.
1.146	0	0	0	0	1	0	0	0	0	0	5.00
1.151	0	0	0	0	0.85	0	0	0.15	0	0	5.36
1.155	0	0	0	0	0.71	0	0.02	0.28	0	0	5.86
1.160	0	0	0	0	0.56	0	0.03	0.41	0	0	6.30
1.164	0	0	0	0	0.41	0	0.04	0.55	0	0	6.73
1.168	0	0	0	0	0.26	0	0.05	0.69	0	0	7.16
1.173	0	0	0	0	0.12	0	0.06	0.83	0	0	7.60
1.177	0	0	0	0	0	0	0	0.92	0.08	0	8.08
1.182	0	0	0	0	0	0	0	0.09	0.91	0	8.91
1.186	0	0	0	0	0	0	0	0	0	1	10.00

Both models give hence the same result. A possible explanation is that in both models, short term rates are more volatile than long term rates, i.e. rolling over is a risky strategy.[292] Furthermore, in these examples the expected value of rolling over once[293] is not very different from investing in longer term bonds, so rollover strategies seem to be disadvantaged. But at least in the Vasicek model, rollover structures are visible in the portfolios.[294] In the HW2 model on the other side no such structures are visible, the optimum portfolio consists only of bonds with a maturity of at least 5 years. Another important

[292] So less weight is given to short bonds.

[293] The model is constructed in such a way that there is only one roll-over, the proceeds from maturing bonds is invested until the investment horizon.

[294] Zero-coupon bonds with maturities less than T are bought.

point is that there are no parallel movements in the Vasicek model. When interest rates are expected to rise, the change in the 10-year spot rate is smaller than the change in the 2-year spot rate, so the expected capital loss on long bonds is relatively smaller than in the duration framework.

Furthermore, we observe that the overall structure of the portfolios is nearly the same regardless of the interest rate scenario considered.[295] This might be due to the fact, that the expected rise in interest rates is quite small.

In summary, we can conclude that the classical duration advice cannot be reproduced in a rising interest rate scenario. This might point to the missing consideration of risk inside the duration framework or to oversimplification (parallel yield-curve shifts).

Yield Curve Strategies

Yield curve strategies involve positioning a portfolio to capitalize on expected changes in the shape of the (risk-free) yield curve.[296] Usually one distinguishes three yield curve strategies: bullet, barbell and ladder strategies.[297] In a bullet strategy, the portfolio is constructed in such a way that the maturity of the securities in the portfolio are highly concentrated at one (intermediate) point on the yield curve.[298]

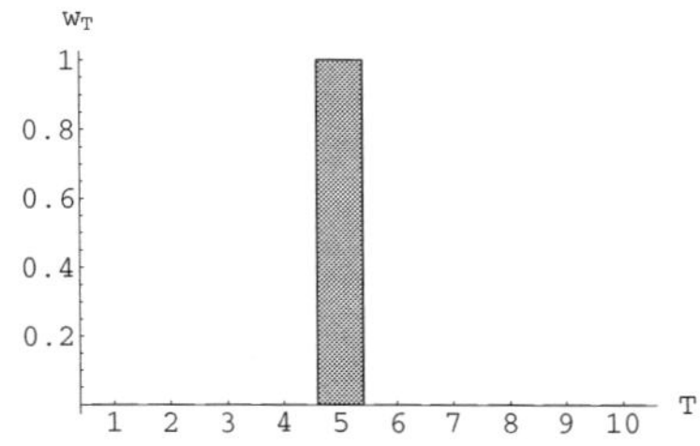

Fig. 4.6. Archetypical bullet portfolio

A barbell portfolio is constructed by concentrating investments at the short-term and the long-term ends of the yield curve.[299] A barbell is defined relative to a bullet strategy, it consists of maturities smaller and greater than the bullet maturity.[300]

[295] Compare Tables 4.8 and 4.19 for the Vasicek model and 4.14 and 4.21 for the HW2 model.
[296] Fabozzi (2004), p. 424.
[297] Fabozzi (2004), p. 427.
[298] Fabozzi (2004), p. 427.
[299] Martellini/Priaulet/Priaulet (2003), p. 239.
[300] Fabozzi (2004), p. 427.

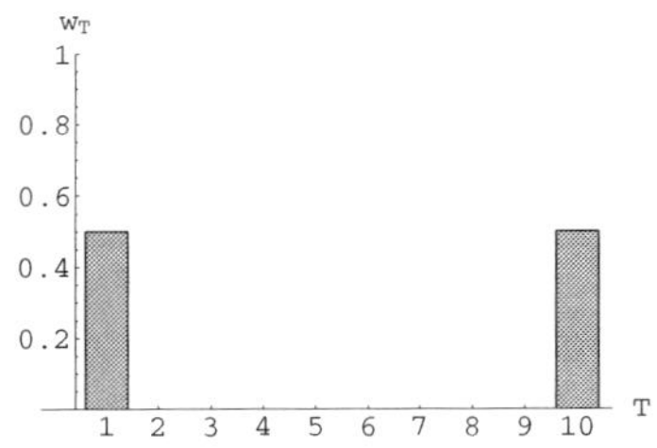

Fig. 4.7. Archetypical barbell portfolio

In a ladder strategy, the portfolio is constructed to have approximately equal amounts of each maturity.[301]

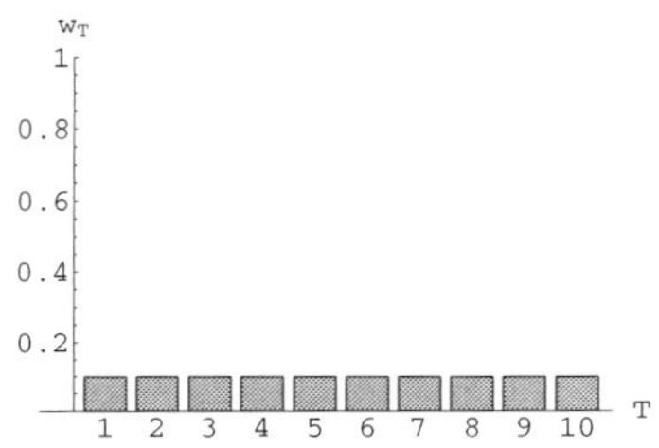

Fig. 4.8. Archetypical ladder portfolio

The fixed income literature dealing with yield curve strategies examined the relative performance of these strategies in different yield curve environments.[302] The main result is that portfolios with equal duration can have quite different risk profiles, i.e. non-parallel changes in the shape of the yield curve can have significant effects on the portfolio return.

Studies showed that a particular yield curve strategy is optimal for a particular interest rate scenario. A flatter yield curve generally[303] results in an outperformance of the barbell strategy (given a downward or upward movement of yield curve) and a steeper yield curve results in an outperformance of the bullet strategy.[304] Fabozzi (2004) finds that for very large level changes[305], a steepening yield curve leads to an outperformance of the barbell strategy – therefore contradicting Jones (1991).[306] Mann/Ramanlal (1997) find that for longer maturities the "normal results" of Jones (1991) are reversed.[307]

[301] Fabozzi (2004), p. 427.
[302] See e.g. Jones (1991), Willner (1996) and Mann/Ramanlal (1997).
[303] The exceptions are considered below.
[304] See Jones (1991), p. 48.
[305] More than $\pm$300bp
[306] See Fabozzi (2004), p. 431.
[307] See Mann/Ramanlal (1997), p. 69.

Next, we want to examine whether bullet, barbell or ladder portfolios result from our mean-variance optimization framework. First, we consider the case of an expected steepening of the term structure and thereafter the case of an expected flattening.

Steepening Term Structure. First, we have a look at the Vasicek model. As we have reiterated in the last section, there are no parallel term structure movements in the Vasicek model. A rising short rate moves the whole term structure upwards (except the infinitely long rate), so the yield spread[308] becomes smaller, i.e. the term structure flattens. On the other hand, a decreasing short rate moves the whole term structure downwards and results in a steeper curve. These two kinds of term structure movements are historically the most common as has been found by Jones (1991). Yield curve strategies are traditionally examined for short-term investment horizons, i.e. in our framework for an investment horizon of 1 year.[309]

We already presented numerical examples for both the Vasicek and the HW2 model, that were set in such an environment. The Vasicek numerical example for the short term investment horizon case that we considered in Chapter 4.2.4, was set in a positive yield curve environment and the term structure was expected to decrease and hence to steepen. The result we obtained[310] was in accordance with the practical advice from yield curve strategies. We expected the term structure to steepen, an environment that favors bullet portfolios.

In the HW2 numerical example in Chapter 4.2.5, the term structure of interest rates was expected to steepen and therefore – according to the practice – bullet portfolios should outperform barbell portfolios. The mean-variance efficient portfolios given in Table 4.12 can be regarded as bullet portfolios[311], therefore the theoretical model comes to the same conclusion as the practical advice for the steepening case.

Flattening Term Structure. The numerical example for the Vasicek model that we presented in Chapter 4.3.2 was set in a positive yield curve environment with an expected rise in the term structure. But the results we obtained were for the long-term investment horizon case (5 years). We use the same parameters and calculate the optimum portfolios for the short-term case. Let $r(0) = 0.02$, then the term structure is expected to increase and therefore to flatten.[312] In such an environment, the practice would propose a barbell portfolio. But as is shown in Table 4.22 the mean-variance approach results in bullet portfolios.

---

[308] Normally defined as the difference between the 10 year spot rate and the 2 year spot rate, we define the yield spread as the difference between the infinitely long rate and the short rate.

[309] See Fabozzi (2004), p. 427.

[310] See Table 4.6.

[311] The investment is usually concentrated among bonds with adjacent maturities.

[312] The other parameters of the term structure model remain the same as in Chapter 4.3.2.

Table 4.22. Vasicek model: Zero-coupon bond weights when a flattening is expected.

$E[W_1]$	w_1	w_2	w_3	w_4	w_5	w_6	w_7	w_8	w_9	w_{10}
1.022	1	0	0	0	0	0	0	0	0	0
1.024	0.44	0.56	0	0	0	0	0	0	0	0
1.025	0	0.85	0.15	0	0	0	0	0	0	0
1.027	0	0.18	0.82	0	0	0	0	0	0	0
1.028	0	0	0.43	0.57	0	0	0	0	0	0
1.030	0	0	0.09	0.38	0.53	0	0	0	0	0
1.031	0	0	0	0	0.59	0.41	0	0	0	0
1.033	0	0	0	0	0	0.41	0.59	0	0	0
1.034	0	0	0	0	0	0	0	0.95	0.05	0
1.036	0	0	0	0	0	0	0	0	0	1

Next, we consider the case of a flattening term structure of interest in the HW2 model.[313] Let $r(0) = 0.015$ and keep the other parameters the unchanged[314], then the term structure of interest rates is expected to increase and to flatten. But as is shown in Table 4.23 the mean-variance approach again results in bullet portfolios.[315]

Table 4.23. HW2 model: Zero-coupon bond weights when a flattening is expected.

$E[W_1]$	w_1	w_2	w_3	w_4	w_5	w_6	w_7	w_8	w_9	w_{10}
1.018	1	0	0	0	0	0	0	0	0	0
1.020	0.78	0	0.22	0	0	0	0	0	0	0
1.021	0.49	0.27	0.09	0.15	0	0	0	0	0	0
1.023	0.34	0	0.66	0	0	0	0	0	0	0
1.024	0.12	0	0.88	0	0	0	0	0	0	0
1.026	0	0	0.67	0.33	0	0	0	0	0	0
1.027	0	0	0	0.93	0.05	0.02	0	0	0	0
1.029	0	0	0	0	0.86	0.14	0	0	0	0
1.030	0	0	0	0	0	0.24	0.76	0	0	0
1.032	0	0	0	0	0	0	0	0	0	1

In our numerical examples, it seems that regardless of the expected twist in the term structure (steepening or flattening) bullet portfolios are mean-variance efficient when short-sales are not allowed.[316] A reason for this might

[313] We cannot use the results from Table 4.21 since this example was set with long-term investment horizon environment.

[314] See page 71.

[315] The initial wealth is spread among bonds with adjacent maturities.

[316] Compare Tables 4.6 and 4.22 for the Vasicek and Tables 4.12 and 4.23 for the HW2 model.

be, that in these term structure models a twist in the term structure (flattening or steepening) can't occur independently of a change in the overall level. Hence, two effects occur at the same time, a change in the level of the term structure and a change in the shape. Nevertheless, the practical advice to buy barbell portfolios when the term structure of interest rates is expected to flatten can't be replicated in our numerical examples.

4.3.3 Passive Bond Portfolio Selection Strategies

Passive portfolio selection is a strategy that requires no – or only minimal – expectational input; typical passive portfolio strategies include indexing and immunization.[317]

Immunization Strategy

The immunization strategy[318] tries to construct a bond portfolio in such a way that the portfolio value at the investment horizon of the investor is immune to a change in interest rates. It is well known that, under the assumptions that the term structure of interest rates is flat and that it changes only in a parallel fashion, this is achieved by constructing a bond portfolio with a duration equal to the investment horizon of the investor.[319] Disregarding the no-short-sales constraint, every possible portfolio duration can be obtained by positions in only two bonds.[320] In this framework, the investor is indifferent between two portfolio with the same duration.

In this section, we compare minimum-variance portfolios[321] in the mean-variance framework described in Chapter 4.2.2 with the duration-based immunization portfolios. The minimum-variance portfolio should – in our opinion – be conceptually as close as it can get to an immunized portfolio. Before we compare the portfolios, we have to adjust the model. The model presented in Chapter 4.2.2 assumes the existence of a riskless zero-coupon bond, i.e. a zero-bond with a maturity equal to the investment horizon of the investor. If we would include this zero-coupon bond, then the minimum-variance portfolio would consist only of the riskless zero-coupon bond and would be identical to the duration immunized portfolio. Hence, for our comparison, we assume

[317] See Fabozzi (2004), p. 412.

[318] Redington (1952) is generally credited with pioneering this strategy.

[319] See Fabozzi (2004), p. 476. The assumption underlying the duration model – a flat term structure that shifts in a parallel fashion – are (i) empirically unlikely and (ii) would represent an arbitrage opportunity. So even if the duration model could been applied to bond portfolio selection, the question remains, whether such an application is sensible.

[320] If short sales aren't allowed, only portfolio durations between the shortest and the longest bond duration are possible but these could still be obtained by positions in just two bonds.

[321] Formally defined in Equation (4.25).

that this riskless zero-coupon bond does not exist. We assume, a bond market where zero-coupon bonds of maturities $1, \ldots, T-1, T+1, \ldots, \tau$ trade. The investment horizon of the investor is T with $1 < T < \tau$. In our numerical example, we set $\tau = 10$. Hence, there are nine risky zero-coupon bonds to invest in. For the sake of analytical tractability, we restrict our analysis again to the two dynamic term structure models we presented in Chapters 3.4 and 3.5, namely the Vasicek (1977) and the HW2 model.

For each term structure model, we consider investment horizons from 2 to 9 years. For each investment horizon, we then calculate the minimum-variance portfolio (subject to short-sale constraints). For comparison, the duration immunization portfolio is constructed by identical weights of the $T-1$ and the $T+1$ zero-coupon bonds.[322] Since we disregard the existence of a riskless zero-coupon bond, we must restate the minimum-variance portfolio optimization problem since it differs from the optimum solution presented in Chapter 4.2.2.

The minimum-variance portfolio is the solution to the following optimization problem

$$\begin{aligned} &\min_{\hat{N}} \ \hat{N}' C \hat{N} \\ &\text{s.t. } \hat{N}' \hat{P}_0 = W_0 \\ &\hat{N} \geq \underline{0} \end{aligned} \tag{4.25}$$

where $\hat{N}$ is the $((\tau-1) \times 1)$ holdings vector, C is the covariance matrix, P_0 is the current price vector, W_0 is the initial wealth and $\underline{0}$ is a $\tau-1$ vector of zeros. Because of the short-sale constraints, this quadratic optimization problem has no analytical solution.[323]

We first analyze the results obtained from the Vasicek model. We choose the same parameters for the Vasicek term structure model as in Chapter 4.2.4.[324] The minimum-variance portfolios for different investment horizons (from 2 to 9) are then given in Table 4.24.

[322] This represents only one possibility of constructing a portfolio with a duration of T.

[323] The numerical calculations are performed with Mathematica 5.2.

[324] See Table 4.2.

Table 4.24. Vasicek model: Portfolio weights of minimum-variance portfolios under short-sale constraints for different investment horizons.

T	w_1	w_2	w_3	w_4	w_5	w_6	w_7	w_8	w_9	w_{10}	MacDur.
2	0.58	0	0.42	0	0	0	0	0	0	0	1.84
3	0	0.54	0	0.46	0	0	0	0	0	0	2.92
4	0	0	0.53	0	0.47	0	0	0	0	0	3.94
5	0	0	0	0.52	0	0.48	0	0	0	0	4.96
6	0	0	0	0	0.51	0	0.49	0	0	0	5.98
7	0	0	0	0	0	0.51	0	0.49	0	0	6.98
8	0	0	0	0	0	0	0.51	0	0.49	0	7.98
9	0	0	0	0	0	0	0	0.51	0	0.49	8.98

It is interesting to note that for every possible investment horizon T only the $T-1$ and the $T+1$ period zero-coupon bonds are included in the bond portfolio. From a risk minimizing perspective it seems that concentrating the maturities of the zero-coupon bonds around the investment horizon is optimal. A look at the correlations confirms this; the chosen zero-coupon bonds are the ones with the minimum correlation of all available assets, e.g. for $T=2$ the minimum correlation is $\text{corr}(\frac{1}{P(1,2)}, P(2,3)) = -0.763153$. The duration immunization portfolios consist of equal weights in the $T-1$ and the $T+1$ period bonds. Therefore, in general, the minimum-variance portfolio puts more weight on the short-term bond. But for larger time-horizons, this difference becomes smaller. Hence, the Macaulay durations of the minimum-variance portfolios are always smaller than the immunized portfolio durations (which are equal to the investment horizon). But, at the end of the day, the portfolio composition is not that important but the mean-variance characteristics of the terminal wealth. Table 4.25 gives the standard deviations of minimum-variance portfolios in the Vasicek model and immunized portfolios, constructed according to the descriptions above.[325]

Table 4.25. Vasicek model: Comparison of minimum-variance portfolio standard deviations to duration-immunized portfolios.

T	$E[W_T]$	Vasicek std.	Duration std.
2	1.03	0.0054	0.0059
3	1.04	0.0049	0.0051
4	1.06	0.0044	0.0045
5	1.08	0.0039	0.0040
6	1.10	0.0034	0.0035
7	1.12	0.0030	0.0031
8	1.14	0.0027	0.0027
9	1.16	0.0024	0.0024

325 The standard deviation for the immunized portfolio has been calculated by $\sqrt{N_d' C N_d}$ where N_d is the holdings vector of the immunized portfolio.

It can be seen that for short-term investment horizons, the portfolio standard deviation differs more than for long-term investment horizons. This is not surprising since for long-term investment horizons, the portfolios are nearly identical. But from a practical perspective we can conclude that the differences are negligible and therefore from a mean-variance perspective, minimum-variance portfolios and duration-immunized portfolios are identical in our example.

We now consider the HW2 model. The one-factor Vasicek model did produce nearly identical results to the duration model – regarding both portfolio weights and the mean-variance profile of terminal wealth. In hindsight this might have been expected, because both models have only one source of randomness. The term structure of interest rates in the Vasicek model doesn't shift in a parallel fashion but still all rates increase or decrease at the same time. The introduction of a second factor results in the possibility of twists (e.g. short-term interest rates increase but long-term interest rates decrease). Therefore a two-factor model may produce significantly different results.

The HW2 minimum-variance portfolios for investment horizons ranging from 2 to 9 years are then given in Table 4.26[326]

Table 4.26. HW2 model: Portfolio weights of minimum-variance portfolios under short-sale constraints for different investment horizons.

T	w_1	w_2	w_3	w_4	w_5	w_6	w_7	w_8	w_9	w_{10}	MacDur.
2	0.72	0	0.28	0	0	0	0	0	0	0	1.56
3	0.21	0.44	0	0.35	0	0	0	0	0	0	2.49
4	0.11	0	0.48	0	0.41	0	0	0	0	0	3.60
5	0.05	0	0	0.50	0	0.45	0	0	0	0	4.75
6	0.03	0	0	0	0.50	0	0.47	0	0	0	5.82
7	0.01	0	0	0	0	0.51	0	0.48	0	0	6.91
8	0	0	0	0	0	0	0.51	0	0.49	0	7.98
9	0	0	0	0	0	0	0	0.51	0	0.49	8.98

In contrast to the one-factor example, the minimum-variance portfolios now usually contain three bonds. In addition to the $T-1$ and $T+1$ period bond, it includes non-negligible positions[327] in the short bond. Hence, the Macaulay portfolio durations of the minimum-variance portfolios differ for short-term investment horizons significantly from the immunized ones, e.g. the 2-period minimum-variance portfolio has a duration of just 1.56 in comparison to a duration of 2.0 for the immunized portfolio. Next, we have again a look at the mean-variance profiles in Table 4.27

[326] Parameter values for the term structure model are given in Table 4.10.

[327] At least for short-term investment horizons.

Table 4.27. HW2 model: Comparison of minimum-variance portfolio standard deviations to duration-immunized portfolios.

T	$E[W_T]$	HW2 std.	Duration std.
2	1.03	0.0055	0.0074
3	1.05	0.0068	0.0081
4	1.06	0.0068	0.0075
5	1.08	0.0061	0.0065
6	1.10	0.0052	0.0054
7	1.12	0.0044	0.0045
8	1.14	0.0036	0.0037
9	1.16	0.0030	0.0030

It can be seen that for short-term investment horizons, the portfolio standard deviation differs more than for long-term investment horizons. Again, this is not surprising since for long-term investment horizons, the portfolios are nearly identical. For shorter investment horizons, there is a non-negligible gain in standard deviation reduction from holding the minimum-variance portfolio instead of the duration immunization portfolio. For all other investment horizons, we can assume that this gain is negligible. But since portfolio immunization strategies are carried out mainly for short-term investment horizons, this still is an important result.

We can conclude that in a one-factor term structure model, the differences between these two strategies are insignificant. Our numerical example for the one-factor Vasicek model showed that the portfolio composition is nearly identical and a possible reduction in portfolio variance when holding the minimum-variance portfolio is negligible. Two-factor interest rate models allow for more realistic term structure movements. In contrast to one-factor models, so-called twists (short-term and long-term rates move in different directions) in the term structure are now possible. But the effect on the immunization performance of the duration portfolio seems to be small. Only for short-term investment horizons is the gain from switching to the minimum-variance portfolio non-negligible. A duration-immunization portfolio hence performs quite well in a two-factor term structure model. When more factors are added, the immunization performance is likely to be poorer.

The comparison outlined above can be linked to studies investigating immunization risk. Fong/Vasicek (1984) explore the impact of portfolio structure on the immunization performance of duration-matched portfolios. In other words, in light of the different ways in which the term structure of interest rates can shift, is it possible to develop a criterion for minimizing the risk that a duration-matched portfolio will not be immunized?[328] We found that from a risk minimizing perspective buying bonds with maturities close to the investment horizon seems optimal.[329] Fong/Vasicek (1984) develop a measure

[328] See Fabozzi (2004), p. 476.

of immunization risk called M-squared. The calculation of M-squared requires only information about the securities under consideration and the investment horizon.[330] According to Bierwag/Fooladi/Roberts (1993) the minimum M-squared portfolio is a bullet portfolio under a specific convexity condition.[331] In our Vasicek example the minimum-variance portfolio had the structure of bullet portfolios as well. But in the HW2 model the structure of the minimum-variance portfolios didn't resemble bullet portfolios. Agca (2002) has comparable findings in that only in some cases minimum M-squared portfolios were also bullet portfolios.[332]

4.3.4 Summary and Conclusion

This part compared real-world bond portfolio selection methods with the mean-variance model. In our numerical examples, most of the time, the results of the mean-variance optimization didn't constitute portfolios that the practice would have suggested in such an interest rate environment. The triggers (or factors) for constructing the portfolio in a specific way from a practitioner's point of view hence seem to play a lesser role in the mean-variance framework. Table 4.28 summarizes the results from this section.[333]

Table 4.28. Comparison with real-world portfolio selection methods.

Strategy	Interest Rates	Model	Equal	Table
Riding	Unchanged	Vasicek	Yes	4.17
Duration	Falling	Vasicek	Yes	4.8
		HW2	Yes	4.14
	Rising	Vasicek	No	4.19
		HW2	No	4.21
Yield Curve Strategies	Steepening	Vasicek	Yes	4.6
		HW2	Yes	4.12
	Flattening	Vasicek	No	4.22
		HW2	No	4.23
Immunization	Falling	Vasicek	Yes	4.24
		HW2	Yes	4.26

When compared to the riding strategy, mean-variance efficient portfolios produced nearly the same results as the proposed real-world portfolios. But the portfolios advocated by the other active strategies could not in every case

329 At least in the Vasicek (1977) model.

330 See Fong/Vasicek (1984), p. 1543.

331 M-squared is a convex function of duration, see Agca (2002).

332 See Agca (2002), p. 74.

333 Column 4 indicates whether the results of the mean-variance optimization coincided with the practical investment advice.

be generated inside the mean-variance framework. An expectation of falling and rising interest rates produced mean-variance efficient portfolios with a Macaulay duration that was greater than the investment horizon. The duration strategy recommends different Macaulay durations for different interest rate scenarios. The recommended portfolios of the yield curve strategies (i.e. bullet or barbell portfolios) could also not be reproduced. A steepening or flattening term structure lead to mean-variance efficient portfolios that resembled bullet portfolios.

The immunization strategy on the other hand, generated portfolios that differed only insignificantly from the mean-variance efficient portfolio, especially in the Vasicek model. From a practical point of view, the duration immunized portfolios perform also quite well when we consider more reasonable term structure movements. This is surely due to the fact, that the level factor in a principal component analysis of the term structure explains most of the variability in interest rates.[334]

A suggestion for future research is the analysis of the performance of mean-variance efficient portfolios and active bond portfolio selection techniques under different scenarios. We suspect that mean-variance efficient portfolios would turn out to be superior, because in this framework the uncertainty regarding the outcome is already being accounted for. One already constructs portfolios in a way that a different outcome than was expected doesn't have a devastating influence on the portfolio. Active bond portfolio strategies on the other hand assume that a particular interest rate scenario is going to materialize.

Another important passive portfolio selection technique is index tracking. It calls for constructing a portfolio in such a way as to match a benchmark portfolio as closely as possible.[335] In practice the benchmark is often a well-known index.[336] For a comprehensive introduction to bond index funds see Mossavar-Rahmani (1991) and for a critical discussion see Granito (1987). In such a framework the appropriate measure of risk is not the standard deviation of portfolio returns but the so-called active return or tracking error.[337] Hence, the objective of the investor is to minimize the tracking error over the next period by adjusting the portfolio weights relative to the benchmark weights. Since this period is generally quite small (could be a day, or month), we do not need a reinvestment assumption.

The general optimization problem is then[338]

[334] See e.g. the seminal study by Litterman/Scheinkman (1991).
[335] Fabozzi (2004), p. 452.
[336] For more information about bond index construction see Brown (1994).
[337] See Fabozzi (2004), p. 415.
[338] See Martellini/Priaulet/Priaulet (2003), p. 218.

$$\min_{w} \operatorname{var}(R_P - R_B)$$
$$\text{s.t. } w' \cdot \underline{1} = 1$$
$$w \geq 0$$

where R_P is the return of the portfolio, R_B is the return of the benchmark, w is the portfolio weights vector and $\underline{1}$ is an appropriate vector of ones. The variance can be written as

$$\bar{w}' C \bar{w}$$

where C is the covariance matrix of bond returns and $\bar{w}$ is the active portfolio, i.e. $\bar{w} = w_B - w$. As becomes obvious from the above formulation, when we set $w = w_B$, then the tracking error becomes zero.[339] In reality, the existence of transaction costs usually forbids perfect replication. In contrast to stock indices, bond indices contain a very large number of bonds and furthermore the index changes quite often due to the exclusion of maturing bonds.[340] Hence, a possible objective is to replicate a bond index consisting of 5.000 bonds with 100 bonds as good as possible thereby introducing an additional constraint to the above formulation of the tracking error minimization problem.

The critical ingredient of any tracking error model is the covariance matrix of returns. Martellini/Priaulet/Priaulet (2003) propose to use the sample covariance matrix of bond returns[341] or a risk factor-based covariance matrix.[342] We suggest using a dynamic term structure model to generate the covariance matrix of bond prices, as has been shown already in this chapter.

A full derivation and application of such a model is outside the scope of this thesis. But a possible line of future research is to compare the tracking error differences resulting from different covariance-generating methods.

---

339 This is called replication, see Martellini/Priaulet/Priaulet (2003), p. 213.

340 See Elton et al. (2003), p. 682.

341 See Martellini/Priaulet/Priaulet (2003), p. 217.

342 See Martellini/Priaulet/Priaulet (2003), p. 222.

5

Dynamic Bond Portfolio Optimization in Continuous Time

5.1 Introduction

The single-period model for bond portfolio optimization is an extreme simplification of reality. It remains nevertheless quite popular in real-life applications because it demands no special knowledge beyond basis probability theory.[343] To overcome the obvious limitations of the single-period model, continuous-time models of asset allocation were developed in the 1970s.[344]

In a continuous-time portfolio selection problem, the investor maximizes his expected utility of terminal wealth (and maybe expected utility of consumption during a specific time interval) by selecting an optimal portfolio strategy, i.e. he chooses the number of assets i to hold at each time t (and in every state of the world). Generally it is assumed that the investor can allocate his wealth to a (locally) riskless money market account and different risky assets. In our setting the risky assets are zero-coupon bonds of different maturities.

The solution of optimal portfolio selection and consumption strategies in continuous time dates back to Merton (1969) and Merton (1971). He was the first to apply the stochastic control (continuous-time equivalent of the dynamic programming) approach[345] to a continuous-time optimal consumption/investment problem.[346] Merton assumed that the interest rate is constant. In order to apply his solution technique to bond portfolio selection problems, the constant interest rate assumption must be dropped.

The stochastic control approach computes the optimal solution to the portfolio selection problem by solving the Hamilton/Jacobi/Bellman (HJB) equa-

343 See Korn (1997), p. 1.

344 The basic continuous-time setting is developed for example in Korn (1997), Munk (2004a), pp. 33 ff. or Björk (1998), pp. 198 ff.

345 Another popular approach for solving the continuous-time asset allocation problem is the martingale approach introduced by Pliska (1986) and Cox/Huang (1989).

346 Munk (2004a), p. 36.

tion[347] in two steps: the first step entails computing the optimal strategy as a function of the unknown optimal expected utility and the second step consists of inserting this strategy into the HJB equation.[348] In order to apply the stochastic control approach, we must assume the existence of a finite-dimensional Markov state process x_t such that the indirect utility function[349] J can be written as $J(t, W_t, x_t)$, where W_t is the time t wealth.[350] In contrast to the martingale approach, one doesn't have to assume a complete market.[351]

There exists a growing literature on bond portfolio selection in continuous time. The following table summarizes the existing literature.[352]

Table 5.1. Existing literature on continuous-time bond portfolio selection.

Model	Term Structure Models	Utility functions	Approach
Sørensen (1999)	Vasicek (1977)	$u(W) = \frac{W^{1-\gamma}-1}{1-\gamma}$	MA
Korn/Kraft (2002)	Ho/Lee (1986), Vasicek (1977)	$u(W) = W^{\gamma}$	SCA
Munk/Sørensen (2004)	Heath/Jarrow/Morton (1992)	$u(W) = e^{\beta T} \frac{W^{1-\gamma}-1}{1-\gamma}$	MA
Kraft (2004)	Ho/Lee (1986), Vasicek (1977), Dothan (1978), Black/Karasinksi (1991), Cox/Ingersoll/ Ross (1985)	$u(W) = W^{\gamma}$	SCA

In this section we want to derive the solution to the bond portfolio selection model for the Vasicek model as in Korn/Kraft (2002) and want to add to the literature by deriving an explicit solution for the HW2 model using the stochastic control approach. In the last part of this chapter, we look at an international bond portfolio selection problem and derive an explicit solution for a two-country problem.

---

347 Will be introduced later in this chapter.

348 See Korn (1997), p. 37.

349 This function will be properly derived later.

350 Munk (2004a), p. 40.

351 See Korn (1997), p. 37.

352 In column 2 the examined term structure model for modeling the bond markets are given, column 3 gives the utility function of the investor and the last column indicates the approach used (MA = martingale approach, SCA = stochastic control approach).

5.2 Bond Portfolio Selection Problem in a HJM Framework

The Heath/Jarrow/Morton (1992) framework introduced in Chapter 3.3 is the most general term structure model. The problem of optimal consumption and investment strategies within a general HJM-framework was addressed by Munk/Sørensen (2004) using the martingale approach.[353] In order to use the traditional stochastic control approach, we restrict our attention to a less general term structure model.

5.2.1 Dynamics of Prices and Wealth

We assume that the shifts in the investment opportunities are (only) driven by a k-dimensional Markov diffusion process x governed by the SDE

$$dx(t) = \alpha(x,t)dt + \beta(x,t)dz(t) \tag{5.1}$$

where $\alpha(x,t)$ is the $(k \times 1)$ drift vector, $\beta(x,t)$ is the $(k \times d)$ matrix of volatilities[354] and $dz(t)$ is a $(d \times 1)$ vector of Brownian motions. This ensures that we can find a discrete set of variables that drives the shifts in the investment opportunity set and satisfies the Markov property.

The investor can choose among n zero-coupon bonds of different maturities and a locally riskfree money market account. He can hence invest his funds in $q = n + 1$ different assets. First, we derive the joint dynamics of the zero-coupon bond prices. With the zero-coupon bond price dynamics from Equation (3.5) and the arbitrage-free condition from Equation (3.13), we can write the dynamics of the n zero-coupon bond prices in matrix notation as follows

$$dP_t = \text{diag}(P_t)\left((r_t 1_n + \sigma_t \lambda_t)dt - \sigma_t dz(t)\right) \tag{5.2}$$

where P_t designates the $(n \times 1)$ vector of zero-coupon bond prices at time t, $\text{diag}(P_t)$ is a $(n \times n)$-diagonal matrix of zero-coupon bond prices, r_t is the value of the short rate at time t, λ_t is a $(d \times 1)$ vector of market prices of risk, 1_n is a $(n \times 1)$ vector of ones and σ_t is a $(n \times d)$ matrix of volatilities.[355]

---

[353] See Table 5.1.

[354] $\beta_{i,j}$ is the volatility of the i-th state variable with respect to the j-th Brownian motion.

[355] The matrix σ_t is hence constructed as follows:

$$\sigma_t = \begin{pmatrix} \sigma_1(t,T_1) & \dots & \sigma_d(t,T_1) \\ \vdots & \vdots & \vdots \\ \sigma_1(t,T_n) & \dots & \sigma_d(t,T_n) \end{pmatrix}$$

The SDE for the money market account was already determined in equation (3.3). The money market account (locally riskless asset) is hence governed by

$$dB_t = B_t r_t dt \tag{5.3}$$

The investor's wealth at time t equals the sum of his portfolio holdings in the q different assets

$$W(t) = N_t' \begin{pmatrix} B_t \\ P_t \end{pmatrix}$$

N_t is the $(q \times 1)$ holdings vector at time t. P is the $(n \times 1)$ vector of zero-coupon bond prices at time t and B is the price of the money market account. Using Itô's Lemma, we obtain the wealth dynamics

$$dW(t) = N_t' \begin{pmatrix} dB_t \\ dP_t \end{pmatrix} + dN_t' \begin{pmatrix} B_t + dB_t \\ P_t + dP_t \end{pmatrix}$$

The first term describes the change in the portfolio value due to changes in prices. The second term describes the change in portfolio value due to changes in the portfolio holdings (rebalancing). We restrict our attention to self-financing portfolios, hence the second term must be equal to zero.[356] Hence, we obtain the following equation for the wealth dynamics

$$dW(t) = N_t' \begin{pmatrix} dB_t \\ dP_t \end{pmatrix}$$

It is advantageous to replace the quantities vector N_t with a formulation containing portfolio weights. We introduce the vector of portfolio weights $\bar{w}$[357]

$$\bar{w} = (w_B, w')'$$

where the scalar w_B denotes the portfolio weight of the money market account and the $(n \times 1)$ vector w contains the portfolio weights of the n zero-coupon bonds. There exists the following relationship between numbers of bonds and fractions of wealth

$$\bar{w}' = \frac{1}{W(t)} N_t' \begin{pmatrix} B_t & 0 \\ 0 & \operatorname{diag}(P_t) \end{pmatrix}$$

We solve for N_t' and obtain

$$N_t' = \bar{w}' W \begin{pmatrix} B_t^{-1} & 0 \\ 0 & \operatorname{diag}(P_t)^{-1} \end{pmatrix}$$

Inserting N_t' in $dW(t)$ we obtain

356 The investor on the one hand does not receive exogenous income and on the other hand doesn't consume. The change in wealth during an infinitesimal time interval is therefore caused only by changes in prices.

357 For notational convenience we ignore the time-dependence of $\bar{w}$.

$$
\begin{aligned}
dW(t) &= \bar{w}'W(t) \begin{pmatrix} \frac{1}{B_t} & 0 \\ 0 & \mathrm{diag}(P_t)^{-1} \end{pmatrix} \begin{pmatrix} dB_t \\ dP_t \end{pmatrix} \\
&= \bar{w}'W(t) \begin{pmatrix} \frac{1}{B_t} & 0 \\ 0 & \mathrm{diag}(P_t)^{-1} \end{pmatrix} \begin{pmatrix} B_t r_t dt \\ \mathrm{diag}(P_t)(r_t 1_n + \sigma_t \lambda_t)dt - \sigma_t dz_t \end{pmatrix} \\
&= \bar{w}'W(t) \begin{pmatrix} r_t dt \\ (r_t 1_n + \sigma_t \lambda_t)dt - \sigma_t dz_t \end{pmatrix} \\
&= (w_B, w) \begin{pmatrix} r_t dt \\ (r_t 1_n + \sigma_t \lambda_t)dt - \sigma_t dz_t \end{pmatrix} W(t) \\
&= W(t) w_B r_t dt + W(t) w'((r_1 1_n + \sigma_t \lambda_t)dt - \sigma_t dz_t)
\end{aligned}
$$

Since portfolio weights must sum to one, we can eliminate w_B using the following equation

$$w_B = 1 - w' 1_n$$

We insert w_B into $dW(t)$, simplify and obtain the dynamics of the investor's wealth

$$dW(t) = W(t)\left((r_t + w_t'\sigma_t\lambda_t)dt - w_t'\sigma_t dz_t\right) \tag{5.4}$$

with $W(0) = W_0$. This equation can be interpreted as a controlled SDE with the control being the portfolio process w.[358]

5.2.2 The Hamilton/Jacobi/Bellman Equation

In this setting, the investor chooses a portfolio process to maximize his expected utility of terminal wealth.

$$\max_{w} E_t\left[u(W(T))\right] \tag{5.5}$$

We define the optimal value function[359] $J(W(t), x_t, t)$

$$J(W(t), x_t, t) \equiv \max_{w} E_t\left[u(W(T))\right] \tag{5.6}$$

Our derivation of the Hamilton/Jacobi/Bellman equation follows Ingersoll (1987) and Kamien/Schwartz (1981).[360] With one period Δt to the investment horizon, the investor solves the problem

$$J(W(T - \Delta t), x(T - \Delta t), T - \Delta t) = \max_{w} E_{T-\Delta t}\left[J(W(T), x_T, T)\right] \tag{5.7}$$

[358] Korn/Kraft (2002), p. 1252.

[359] Also referred to as the indirect (or derived) utility function. Since we are operating in a market with stochastic investment opportunities the function J depends – besides t and W – on the vector of state variables x.

[360] See Ingersoll (1987), pp. 271–274 and Kamien/Schwartz (1981), pp. 246–247.

Continuing in a recursive fashion, we eventually obtain[361]

$$J(W(t), x_t, t) = \max_{w} E_t\left[J(W_t + \Delta W_t, x_t + \Delta x_t, t + \Delta t)\right] \tag{5.8}$$

We do a Taylor series expansion of J around $(t, W, x_1, \ldots, x_d)$ to get[362]

$$\begin{aligned}
&J(W_t + \Delta W_t, x_1 + \Delta x_1, \ldots, x_d + \Delta x_d, t + \Delta t) = \\
&J(t, W_t, x_1, \ldots, x_d) + J_t \Delta t + J_W \Delta W + \sum_{i=1}^{d} J_{x_i} \Delta x_i + \sum_{i=1}^{d} J_{x_i W} \Delta x_i \Delta W \\
&+ \frac{1}{2} J_{WW} \Delta W^2 + \frac{1}{2} \sum_{i=1}^{d} \sum_{j=1}^{d} J_{x_i x_j} \Delta x_i \Delta x_j
\end{aligned} \tag{5.9}$$

This result can be written more readable in matrix notation[363]

$$\begin{aligned}
&J(W_t + \Delta W_t, x + \Delta x, t + \Delta t) = \\
&\quad J(t, W_t, x_t) + J_t \Delta t + J_W \Delta W + J_x' \Delta x + J_{xW}' (\Delta x)(\Delta W) \\
&\quad + \frac{1}{2} J_{WW} (\Delta W)^2 + \frac{1}{2} \operatorname{tr}(J_{xx'} (\Delta x)(\Delta x'))
\end{aligned}$$

where

$$J_x = \begin{pmatrix} J_{x_1} \\ \vdots \\ J_{x_d} \end{pmatrix}, J_{xW} = \begin{pmatrix} J_{x_1 W} \\ \vdots \\ J_{x_d W} \end{pmatrix}, J_{xx'} = \begin{pmatrix} J_{x_1 x_1} & \ldots & J_{x_1 x_d} \\ \vdots & & \vdots \\ J_{x_d x_1} & \ldots & J_{x_d x_d} \end{pmatrix}$$

We calculate the expectation

$$\begin{aligned}
&E[J(W + \Delta W, x + \Delta x, t + \Delta t)] = \\
&\quad J(W, x, t) + J_t \Delta t + J_W E[\Delta W] + J_x E[\Delta x] + J_{xW} E[\Delta x \Delta W] \\
&\quad + \frac{1}{2} J_{WW} E[\Delta W^2] + \frac{1}{2} \operatorname{tr}(J_{xx'} E[\Delta x \Delta x'])
\end{aligned}$$

and insert this expression into Equation (5.8). We obtain

$$\begin{aligned}
0 = \max_{w} \Big(&J_t \Delta t + J_W E[\Delta W] + J_x' E[\Delta x] + J_{xW}' E[\Delta x \Delta W] \\
&+ \frac{1}{2} J_{WW} E[\Delta W^2] + \frac{1}{2} \operatorname{tr}(J_{xx'} E[\Delta x \Delta x']) \Big)
\end{aligned} \tag{5.10}$$

Now we calculate the necessary expectations. With Δx from (5.1) and ΔW from (5.4) we obtain[364]

---

[361] See Ingersoll (1987), p. 273.

[362] Apply the following multiplication rules for Brownian motions $\Delta t^2 = 0$, $\Delta t \Delta x_i = 0$ and $\Delta t \Delta W = 0$, see Malliaris/Brock (1982), p. 87.

[363] $\operatorname{tr}(M)$ designated the trace of the matrix M.

[364] For a derivation see Appendix B.

$$\begin{aligned} E[\Delta x] &= \alpha \Delta t \\ E[\Delta W] &= (r_t + w'\sigma\lambda)W\Delta t \\ E[\Delta x \Delta x'] &= \beta\beta'\Delta t \\ E[\Delta x \Delta W] &= -\beta\sigma' w W \Delta t \\ E[\Delta W^2] &= (w'\sigma\sigma' w)W^2 \Delta t \end{aligned}$$

We insert these expressions into (5.10) and obtain

$$\begin{aligned} 0 = \max_w \big(& J_t \Delta t + J_W (r_t + w^T \sigma\lambda) W \Delta t + J_x' \alpha \Delta t - J_{xW}' (\beta\sigma' w) W \Delta t \\ & + \frac{1}{2} J_{WW} (w'\sigma\sigma' w) W^2 \Delta t + \frac{1}{2} \mathrm{tr}(J_{xx'} \beta\beta') \Delta t \Big) \end{aligned}$$

We divide by Δt, let $\Delta t \to 0$ and eventually obtain the HJB equation[365]

$$\begin{aligned} 0 = \max_w \big(& J_t + J_W (r_t + w^T \sigma\lambda) W + J_x' m - J_{xW}' (s\sigma' w) W \\ & + \frac{1}{2} J_{WW} (w'\sigma\sigma' w) W^2 + \frac{1}{2} \mathrm{tr}(J_{xx'} s s') \Big) \end{aligned} \tag{5.11}$$

with boundary condition

$$\begin{aligned} J(W(T), x_T, T) &= \max_w \; E_T[u(W(T))] \\ &= u(W(T)) \end{aligned} \tag{5.12}$$

5.2.3 Derivation of Optimum Portfolio Weights

For notational convenience Equation (5.11) can be written as follows

$$0 = \max_w (\phi(w; W, x, t)) \tag{5.13}$$

We get the n first-order conditions for an optimum by calculating the derivative of ϕ w.r.t. w

$$\phi_w = W\sigma\lambda J_W - W\sigma s' J_{xW} + \sigma\sigma' w^* W^2 J_{WW} = \underline{0} \tag{5.14}$$

where $\underline{0}$ is a $(n \times 1)$ vector of zeros. Solving for w^* we obtain

$$w^* = \left(-\frac{J_W}{W J_{WW}} \right) (\sigma\sigma')^{-1} \sigma\lambda + (\sigma\sigma')^{-1} \sigma\beta' \frac{J_{xW}}{W J_{WW}}$$

We define $\Sigma \equiv \sigma\sigma'$ and $\hat{\Sigma} \equiv \sigma\beta'$ and can the solution in the following shorter form

$$w^* = \left(-\frac{J_W}{W J_{WW}} \right) \Sigma^{-1} \sigma\lambda + \Sigma^{-1} \hat{\Sigma} \frac{J_{xW}}{W J_{WW}} \tag{5.15}$$

[365] See Ingersoll (1987), p. 282.

The necessary conditions are[366]

$$\phi_{ww} < 0$$
$$\det(\phi_{ww}) > 0$$

Since

$$\phi_{ww} = W^2 J_{WW} \Sigma$$

and Σ is positive-definite, we have a single necessary condition for an interior maximum, namely $J_{WW} < 0$.[367]

Number of Zero-Coupon Bonds to Include in Bond Portfolio

At the beginning of this section, we simply stated that the investor can invest in n zero-coupon bonds of different maturity and in a money market account. We didn't elaborate on how many different bonds the investor really needs to consider. That depends on the interest rate model under consideration and is closely related to the question of duplicable cash flows. The optimum portfolio weights from Equation (5.15) are

$$w^* = \left(-\frac{J_W}{W J_{WW}}\right) \Sigma^{-1} \sigma \lambda + \Sigma^{-1} \hat{\Sigma} \frac{J_{xW}}{W J_{WW}}$$

This equation has a solution if the inverse of Σ exists. The matrix Σ can be written as follows

$$\Sigma = \begin{pmatrix} \sum_{i=1}^d \sigma_i(t,T_1)^2 & \sum_{i=1}^d \sigma_i(t,T_1)\sigma_i(t,T_2) & \dots & \sum_{i=1}^d \sigma_i(t,T_1)\sigma_i(t,T_n) \\ \vdots & \vdots & & \vdots \\ \sum_{i=1}^d \sigma_i(t,T_n)\sigma_i(t,T_1) & \sum_{i=1}^d \sigma_i(t,T_n)\sigma_i(t,T_2) & \dots & \sum_{i=1}^d \sigma_i(t,T_n)^2 \end{pmatrix}$$

This is the symmetric covariance matrix of instantaneous zero-coupon bond returns

$$\Sigma = \begin{pmatrix} \operatorname{var}\left(\frac{dP(t,T_1)}{P(t,T_1)}\right) & \operatorname{cov}\left(\frac{dP(t,T_1)}{P(t,T_1)}, \frac{dP(t,T_2)}{P(t,T_2)}\right) & \dots & \operatorname{cov}\left(\frac{dP(t,T_1)}{P(t,T_1)}, \frac{dP(t,T_n)}{P(t,T_n)}\right) \\ \vdots & & & \vdots \\ \operatorname{cov}\left(\frac{dP(t,T_n)}{P(t,T_n)}, \frac{dP(t,T_1)}{P(t,T_1)}\right) & \operatorname{cov}\left(\frac{dP(t,T_n)}{P(t,T_n)}, \frac{dP(t,T_2)}{P(t,T_2)}\right) & \dots & \operatorname{var}\left(\frac{dP(t,T_n)}{P(t,T_n)}\right) \end{pmatrix}$$

The inverse of a quadratic matrix exists if the matrix has a determinant that is not zero.[368] The determinant of a matrix is zero if[369]

[366] See Merton (1992), p. 129.
[367] See Merton (1992), p. 129.
[368] See Bronstein et al. (1999), p. 255.
[369] See Bronstein et al. (1999), p. 259.

1. One row consists only of zeros.
2. Two rows are identical.
3. One row is a linear combination of other rows.

We now consider each of the different cases. Case 1 can only occur, if we include a (locally) riskless asset in the zero-coupon bond set. A (locally) riskless asset is a zero-coupon bond with instantaneous maturity. This bond is (economically) equivalent to the money market account. The money market account is considered separately and is therefore not included in the covariance matrix of instantaneous zero-coupon bond returns. Hence case 1 is precluded.

The occurrence of two identical rows in the covariance matrix happens to coincide with the situation that two assets are identical. Two zero-coupon bonds are identical if they have the same maturity. This is also precluded.

Last but not least we consider case 3. Now the concept of stochastic linear dependence is useful. It is well known that there can be only so many stochastic linear independent assets as there are risk factors (Brownian motions). In our framework, there are d Brownian motions that drive the underlying uncertainty. Therefore, there can be a maximum of d stochastic linear independent assets. Hence, the covariance matrix is invertible if the zero-coupon bonds included in the investment opportunity set consist of a maximum of d zero-coupon bonds with different maturity.

In reality, more than d zero-coupon bonds trade in the market[370] hence we can choose a subset of these traded bonds. Economically this means, that d different zero-coupon bonds and a position in the money market account can dynamically replicate all interest rate dependent securities (and hence all other zero-coupon bonds).

Economic Interpretation of Optimum Portfolio Weights

The optimum portfolio weights in Equation (5.15) have been subject to far-ranging analysis from different authors. We first summarize these findings and then look for differences between the equity market interpretations and the bond market.

The optimum portfolio weights consist of a mean-variance-efficient portfolio and k hedge portfolios.[371] The fact that

$$\left(-\frac{J_W}{WJ_{WW}}\right)\Sigma^{-1}\sigma\lambda$$

is mean-variance efficient and $\Sigma^{-1}\sigma\lambda$ is the Tobin-Fund has been shown by Ingersoll (1987).[372] Since J depends among other things on the time from t

[370] Even in a multi-factor term structure model d is usually quite small in accordance with empirical research from e.g. Litterman/Scheinkman (1991). They concluded that d should be no greater than 3, i.e. three factors account for nearly 100 % of the variability in interest rates.

[371] See Nietert (1996), p. 54.

[372] For a derivation see Ingersoll (1987), p. 283.

to T, a simultaneous optimization takes place and no sequential ("myopic") decision making.[373]

In this general model, the investment opportunity set is stochastic, i.e. changes in the state variables change the nature of the available assets. Ingersoll (1987) finds that the hedge portfolios[374] provides the best possible hedge against change in the state variables since it has the maximum absolute correlation with it.[375] We introduce the following notation. Let $\hat{\Sigma}_i$ denote the i-th column of $\hat{\Sigma}$, then we can write (5.15) as follows[376]

$$w^* = \left(-\frac{J_W}{WJ_{WW}}\right)\Sigma^{-1}\sigma\lambda + \sum_{i=1}^{k}\Sigma^{-1}\hat{\Sigma}_i\frac{J_{x_iW}}{WJ_{WW}}$$

where

$$\hat{\Sigma}_i = \left(\mathrm{var}\left(\frac{dP(t,T_1)}{P(t,T_1)}, dx_i\right), \ldots, \mathrm{var}\left(\frac{dP(t,T_n)}{P(t,T_n)}, dx_i\right)\right)'$$

A hedge portfolio for a state variable can hence only be constructed if the state variable is correlated with at least one asset.[377] For a thorough analysis of the hedge portfolios in a two asset and one state variable case see Nietert (1996), pp. 55–59. We have a further look at the economic interpretation of the optimum portfolios in a later section.

5.2.4 The Value Function for CRRA Utility Functions

The optimum portfolio weights in Equation (5.15) still depend on the unknown function J. Finding a solution for J requires the insertion of the optimum weights in the HJB, we obtain

$$\begin{aligned} J_t + J_W rW + J_x'\alpha + \frac{1}{2}\mathrm{tr}(J_{xx'}\beta\beta') - \frac{1}{2}\frac{J_W^2}{J_{WW}}(\sigma\lambda)'(\sigma\sigma')^{-1}(\sigma\lambda) \\ + \frac{J_W}{J_{WW}}J_{xW}'\beta\sigma'(\sigma\sigma')^{-1}\sigma\lambda - \frac{1}{2}\frac{1}{J_{WW}}J_{xW}'\beta\sigma'(\sigma\sigma')^{-1}\sigma\beta' J_{xW} = 0 \end{aligned} \tag{5.16}$$

Simplification yields[378]

$$\begin{aligned} J_t + J_W rW + J_x'\alpha + \frac{1}{2}\mathrm{tr}(J_{xx'}\beta\beta') - \frac{1}{2}\frac{J_W^2}{J_{WW}}(\sigma\lambda)'(\sigma\sigma')^{-1}(\sigma\lambda) \\ + \frac{J_W}{J_{WW}}J_{xW}'\beta\lambda - \frac{1}{2}\frac{1}{J_{WW}}J_{xW}'\beta\beta' J_{xW} = 0 \end{aligned} \tag{5.17}$$

---

373 See Nietert (1996), p. 20.

374 $\Sigma^{-1}\hat{\Sigma}\frac{J_{xW}}{WJ_{WW}}$

375 See Ingersoll (1987), p. 282.

376 See Nietert (1996), p. 54.

377 See Nietert (1996), p. 54.

378 $\sigma(\sigma\sigma')^{-1}\sigma = \mathrm{Id}_d$, where Id_d is the $(d \times d)$ identity matrix.

We multiply with J_{WW} and obtain

$$J_t J_{WW} + J_{WW} J_W r W + J_{WW} J_x' \alpha + \frac{1}{2} J_{WW} \mathrm{tr}(J_{xx'} ss') - \frac{1}{2} J_W^2 (\sigma\lambda)'(\sigma\sigma')^{-1}(\sigma\lambda) + J_W J_{xW}' \beta\lambda - \frac{1}{2} J_{xW}' \beta\beta' J_{xW} = 0 \quad (5.18)$$

Unfortunately the HJB is usually highly non-linear and hence explicit solutions are only available in some cases. In accordance with the existing literature on continuous-time bond portfolio selection[379] we assume that the preferences of the investors can be represented by a utility function with constant relative risk aversion (CRRA) of the form

$$u(W(T)) = W^\gamma \quad (5.19)$$

with $0 < \gamma < 1$.[380] We now conjecture that the unknown value function can be separated as follows[381]

$$J(W(t), x_t, t) = f(x,t)u(W(T)) = f(x,t)W^\gamma \quad (5.20)$$

We insert the partial derivatives[382] of $J(W(t), x_t, t)$ into (5.18) and after reducing $W^{2\gamma-2}$ and γ, we obtain a PDE for the unknown function $f(x,t)$

$$(\gamma - 1) f_t f + \gamma(\gamma - 1) f^2 r + (\gamma - 1) f f_x' \alpha + \frac{1}{2}(\gamma - 1) f \mathrm{tr}(f_{xx'} \beta\beta') - \frac{1}{2}\gamma f^2 (\sigma\lambda)'(\sigma\sigma')^{-1}(\sigma\lambda) + \gamma f f_x' \beta\lambda - \frac{1}{2}\gamma f_x' \beta\beta' f_x = 0 \quad (5.21)$$

with boundary condition $f(x_T, T) = 1$. With (5.21) we are in a position to find a solution for the unknown value function $J(W(t), x_t, t)$.

With $J(W(t), x_t, t)$ from Equation (5.20), we obtain the following expression for the optimum portfolio weights in the CRRA case

$$w^* = \frac{1}{1-\gamma}\left(\Sigma^{-1}\sigma\lambda - \Sigma^{-1}\hat{\Sigma}\frac{f_x}{f}\right) \quad (5.22)$$

[379] See Table 5.1.

[380] Korn/Kraft (2002), p. 1252.

[381] Analogous to Korn (1997), p. 52 or Korn/Kraft (2002), p. 1254.

[382] The partial derivatives are

$$\begin{aligned} J_t &= f_t W^\gamma \\ J_x &= f_x W^\gamma \\ J_W &= \gamma W^{\gamma-1} f \\ J_{WW} &= \gamma(\gamma-1) W^{\gamma-2} f \\ J_{xx} &= f_{xx} W^\gamma \\ J_{xW} &= \gamma W^{\gamma-1} f_x \end{aligned}$$

In the next section, we want to derive two special cases. First we want to examine dynamic bond portfolio selection in a Vasicek model and then derive the optimum portfolio weights in a HW2 model.

5.3 Special Cases

5.3.1 One-Factor Vasicek (1977) Model

The dynamic bond portfolio selection problem in a Vasicek (1977) term structure model has already been solved by Korn/Kraft (2002) using the stochastic control approach. We derive their result as a special case of Equation (5.22).

Derivation the Optimum Portfolio Weights

In the Vasicek model the only state variable is the short rate of interest, i.e. $x(t) = r(t)$ and $k = 1$. Furthermore, the short rate is influenced by one Brownian motion only ($d = 1$). As we have shown in Section 5.2.3 the number of zero-coupon bonds that we have to consider depends on the number of Brownian motions. In this case, the optimum portfolio consists of holdings in the riskless money market account and in one risky zero-coupon bond of maturity T_1 ($n = 1$).

With $k = d = n = 1$, Equation (5.22) reduces to

$$w_t^* = \frac{1}{1-\gamma}\left(\frac{\lambda}{\sigma(t,T_1)} - \frac{\beta(t)}{\sigma(t,T_1)}\frac{f_r}{f}\right) \tag{5.23}$$

We specify the function f introduced in Equation (5.20) as follows[383]

$$f(r,t) = g(t)\exp(v(t)r) \tag{5.24}$$

then

$$\frac{f_r}{f} = v(t)$$

and the portfolio weights in Equation (5.22) become

$$w^* = \frac{1}{1-\gamma}\left(\frac{\lambda}{\sigma(t,T_1)} - \frac{\beta(t)}{\sigma(t,T_1)}v(t)\right) \tag{5.25}$$

In the Vasicek model the volatility of the short rate is a constant, i.e. $\beta(t) = \sigma_r$.[384] Furthermore, the volatility of a bond maturing at time T_1 is $\sigma(t,T_1) = \sigma_r B(t,T_1) = \sigma_r\left(\frac{1-e^{-\kappa(T_1-t)}}{\kappa}\right)$.[385]

[383] See Korn/Kraft (2002), p. 1254.
[384] See Equation (3.16).
[385] See Korn/Kraft (2002), p. 1252.

We insert (5.24) in (5.21), set the drift of the short rate equal to[386] $\alpha = \kappa(\theta - r)$ and obtain

$$\begin{aligned}&e^{2rv(t)}r(\gamma-1)g(t)^2\left(\gamma-\kappa v(t)+v'(t)\right)-\frac{1}{2}e^{2rv(t)}g(t)\\&\times\left(g(t)\left(\gamma\lambda^2+\sigma_r^2v(t)^2-2((\gamma-1)\theta\kappa+\sigma_r\gamma\lambda)v(t)\right)\right.\\&\left.-2(\gamma-1)g'(t)\right)=0\end{aligned} \tag{5.26}$$

subject to the boundary conditions $v(T) = 0$ and $g(T) = 1$.[387] Since

$$r(\gamma-1)g(t)^2\left(\gamma-\kappa v(t)+v'(t)\right)=0$$

must hold for all values of r, the term $(\gamma - \kappa v(t) + v'(t))$ must be equal to zero. This is a second order ODE for the unknown function $v(t)$ subject to $v(T) = 0$. The solution is[388]

$$v(t)=\frac{\left(1-e^{-\kappa(T-t)}\right)\gamma}{\kappa} \tag{5.27}$$

It is interesting to note that

$$v(t)=\gamma B(t,T)$$

With $v(t)$ from (5.27), Equation (5.26) becomes an ODE for the unknown function $g(t)$. The function $g(t)$ is not needed for the optimum portfolio strategy, therefore we give its solution only in the Appendix.[389] Hence, $v(t)$ from (5.27) and $g(t)$ from (C.1) solve (5.26). The optimal portfolio strategy can then be written as[390]

$$w_t^*=\frac{1}{1-\gamma}\left(\frac{\lambda\kappa}{\sigma_r(1-\exp(-\kappa(T_1-t)))}-\gamma\frac{1-\exp(-\kappa(T-t))}{1-\exp(-\kappa(T_1-t))}\right)$$

or

$$\begin{aligned}w_t^*=&\underbrace{\frac{1}{1-\gamma}}_{-\frac{J_W}{WJ_{WW}}}\left(\frac{\lambda\kappa}{\sigma_r(1-e^{-\kappa(T_1-t)})}\right)\\&-\underbrace{\frac{\gamma}{1-\gamma}\frac{1-e^{-\kappa(T-t)}}{\kappa}}_{\frac{J_{rW}}{WJ_{WW}}}\left(\frac{\kappa}{1-e^{-\kappa(T_1-t)}}\right)\end{aligned} \tag{5.28}$$

---

386 See Equation (3.16).

387 Since $f(r,T) = 1$.

388 See Korn/Kraft (2002), p. 1255.

389 See Appendix C.1, Equation (C.1).

390 See Korn/Kraft (2002), p. 1255.

Interpretation of Optimum Portfolio Weights

As has been discussed in the last section, the optimum portfolio strategy in Equation (5.28) consists of a mean-variance efficient portfolio and one hedge portfolio. Furthermore, since all parameters are constant and the short rate has no influence on the portfolio weights, it is a deterministic function of time. We first examine the mean-variance efficient portfolio. As can be seen from Equation (5.28) it is independent of the investment horizon and can be regarded as a product of a volume and a structural component.

$$\underbrace{\frac{1}{1-\gamma}}_{\text{volume component}} \underbrace{\left(\frac{\lambda\kappa}{\sigma_r(1-e^{-\kappa(T_1-t)})}\right)}_{\text{structural component}}$$

The volume component depends only on the risk aversion of the investor. It is positive[391] and increasing in γ. The structural component reflects the trade-off between expected risk premium to risk contribution.[392]

Next, we analyze the hedge portfolio. We have stated in Section 5.2.3 that the hedge portfolio has the maximum absolute correlation possible. In a one factor model this is no defining characteristic since the hedge portfolio can consist only of one zero-coupon bond and every zero-coupon bond has the same instantaneous correlation with the state variable (short rate), namely -1.[393] The hedge portfolio consists of a volume term and a structural component as well

$$\underbrace{\frac{\gamma}{\gamma-1}\frac{1-e^{-\kappa(T-t)}}{\kappa}}_{\text{volume component}} \underbrace{\left(\frac{1}{1-e^{-\kappa(T_1-t)}}\right)}_{\text{structural component}}$$

The volume component depends on the risk aversion of the investor, the investment horizon and on some distributional parameters of the state variable (the short rate). It does not depend on the level of volatility σ_r and the market price of interest rate risk. As the investment horizon approaches, the hedge portfolio position in the zero-coupon bond approaches zero since

$$\lim_{t\to T}\frac{\gamma}{\gamma-1}\frac{1-e^{-\kappa(T-t)}}{\kappa}=0$$

This is economically plausible, since changes in the investment opportunity set have a smaller effect on the portfolio value as the investment horizon approaches.

---

391 Since $0<\gamma<1$.

392 See Nietert (1996), p. 20.

393 With dr from (3.16) and the dP from (3.5) one derives for the Vasicek model a correlation coefficient ϱ of -1. Economically speaking bond prices move inversely to interest rates.

For the portfolio as a whole the following results can be derived. The market prices of interest rate risk λ determine the relative attractiveness of the zero-coupon bond relative to the money market account, it is straightforward to show that $\frac{\partial w_t^*}{\partial \lambda} > 0$. The short rate volatility σ_r which determines the volatility level for all other interest rates has a plausible effect too. If the volatility declines (rises) than the optimum portfolio weight of the zero-coupon bond rises (declines), i.e. $\frac{\partial w^*}{\partial \sigma} < 0$ if $\lambda > 0$. The investor has no position in the zero-coupon bond at time t if

$$\frac{1}{\gamma}\frac{\lambda}{\sigma_r} = \frac{1 - e^{-\kappa(T-t)}}{\kappa}$$

Since the left hand side is constant and the right hand side is a function of t, such an occurrence seems to be quite rare.

Numerical Example

The practical application of the model is problematic since it is extremely sensitive to (unobservable) input parameters, i.e. estimation errors have quite a significant effect on the portfolio weights. To illustrate the effect, suppose that $\gamma = 0.5, \sigma = 0.02, \kappa = 0.2, T - t = 5$ and $T_1 - t = 10$, i.e. the investment horizon is five years and the bond under consideration is a 10 year zero-coupon bond. The following table gives the portfolio weight of the risky zero-coupon bond at time t for different values of λ.

Table 5.2. Vasicek model: Zero-coupon bond weight for different market prices of interest rate risk.

λ	0	0.02	0.04	0.06	0.08	0.1	0.12	0.14	0.16	0.18	0.2
w^*	-0.73	-0.27	0.19	0.66	1.12	1.58	2.04	2.51	2.97	3.43	3.9

We assume that $\lambda = 0.04$ and we are now interested in the behavior of the optimum portfolio weight as a function of time, in other words, how does the zero-coupon bond weight change as the investment horizon approaches? We find that the weight of the risky zero-coupon bond increases as we approach the investment horizon.[394] The portfolio weight as a function of time is shown in Figure 5.1.[395]

394 This behavior (i.e. rising risky bond weight as the investment horizon approaches) is not true in general, but only if $e^{T_1\kappa}\gamma\sigma + e^{T\kappa}(\kappa\lambda - \gamma\sigma) > 0$. To arrive at this solution calculate the partial derivative of w^* w.r.t t and let this result be greater or equal to zero and simplify.

395 Furthermore for different values of λ this graph simply moves in parallel, this is obvious since the partial derivative of the optimum portfolio weights w.r.t. λ is constant.

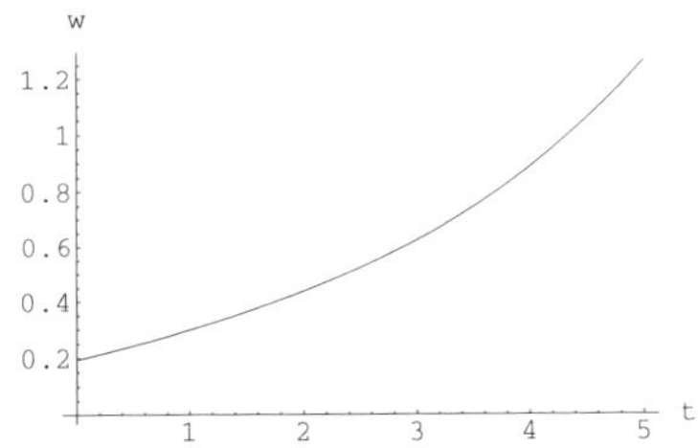

Fig. 5.1. Vasicek Bond Portfolio Selection: Zero-Coupon Bond weight w as a function of time

This suggested investment behavior is contrary to popular investment advice. Typical investment advice is to decrease risky holdings as the investment horizon approaches.

Comparison with Classical Active Bond Portfolio Strategies

In the last chapter on static bond portfolio selection, we compared the resulting portfolios to classical yield curve strategies. We refrain from comparing the dynamic models for several reasons. First, in the dynamic models, the underlying bond market uncertainty determines to a large extent the portfolio structure. Given a one-factor model, we need a position in the money market account and a position in a zero-coupon bond with a maturity greater than the investment horizon. Hence, no ladder portfolios are possible. Second, the portfolio weights in the dynamic models can become negative as can be seen from Table 5.2. The real-world methods we introduced in the last chapter assumed that all portfolio weights are positive. But if we restrict short sales in the dynamic models, then the problem becomes far more difficult to solve.[396]

5.3.2 Two-Factor Hull/White (1994) Model

In this section we want to examine bond portfolio selection in a HW2 model. Munk/Sørensen (2004) solve a dynamic bond portfolio selection problem in the HJM framework using the martingale approach. We derive an explicit solution for the HW2 model using the stochastic control approach and hence add to the existing literature on continuous-time bond portfolio optimization.

Derivation of Optimum Portfolio Weights

The HW2 model has two state variables: the short rate of interest $r(t)$ and the mean reversion level $\varepsilon(t)$, hence the vector of state variables is

[396] Constrained optimization problems in continuous-time have been studied by Cvitanic/Karatzas (1992), Xu/Shreve (1992a) and Xu/Shreve (1992b).

$x(t)' = (r(t), \varepsilon(t))$. The underlying uncertainty is driven by two uncorrelated Brownian motions $z_1(t)$ and $z_2(t)$. Hence – as has been shown in Section (5.2.3) – it is necessary to hold two zero-coupon bonds of maturities T_1 and T_2 and the money market account, i.e. $k = d = n = 2$. We can specify the quantities in the general portfolio weight Equation (5.22) as follows

$$\sigma = \begin{pmatrix} \sigma_1(t,T_1) & \sigma_2(t,T_1) \\ \sigma_1(t,T_2) & \sigma_2(t,T_2) \end{pmatrix}$$
$$\lambda = \begin{pmatrix} \lambda_1 \\ \lambda_2 \end{pmatrix}$$
$$\beta = \begin{pmatrix} \sigma_r & 0 \\ \varrho\sigma_\varepsilon & \sqrt{1-\varrho^2}\sigma_\varepsilon \end{pmatrix}$$
$$f_x = \begin{pmatrix} f_r \\ f_\varepsilon \end{pmatrix}$$

A closed-form solution for the optimal portfolio weights can be found if we assume the following functional form for $f(x,t)$. The function $f(x,t)$ separates the indirect utility function J and has been introduced in Equation (5.20)

$$f(r,\varepsilon,t) = g(t)e^{p(t)r(t)+q(t)\varepsilon(t)}$$

where $p(t)$ and $q(t)$ are deterministic functions of time, hence

$$\frac{f_x}{f} = \frac{1}{f}\begin{pmatrix} f_r \\ f_\varepsilon \end{pmatrix} = \begin{pmatrix} p(t) \\ q(t) \end{pmatrix}$$

The zero-coupon bond volatilities $\sigma_1(t,T)$ and $\sigma_2(t,T)$ are[397]

$$\sigma_1(t,T) = \sigma_r B_1(t,T) + \varrho\sigma_\varepsilon B_2(t,T)$$
$$\sigma_2(t,T) = \sqrt{1-\varrho^2}\sigma_\varepsilon B_2(t,T)$$

Inserting the assumption for f in Equation (5.21) leads to

$$\begin{aligned} &-\frac{e^{rp(t)+\epsilon q(t)}g(t)\left(\gamma\lambda_1^2+\gamma\lambda_2^2+\sigma_r^2 p(t)^2+\sigma_\epsilon^2 q(t)^2\right)}{2(\gamma-1)} + e^{rp(t)+\epsilon q(t)}g(t) \\ &\times\frac{\gamma\left(\varrho\lambda_1+\sqrt{1-\varrho^2}\lambda_2\right)\sigma_\epsilon q(t)+p(t)\left((\gamma-1)\theta+\gamma\lambda_1\sigma_r-\varrho\sigma_r\sigma_\epsilon q(t)\right)}{\gamma-1} \\ &+e^{rp(t)+\epsilon q(t)}g'(t)+e^{rp(t)+\epsilon q(t)}rg(t)\left(\gamma-\kappa_r p(t)+p'(t)\right)+ \\ &+e^{rp(t)+\epsilon q(t)}\epsilon g(t)\left(p(t)-\kappa_\epsilon q(t)+q'(t)\right)=0 \end{aligned} \tag{5.29}$$

It can be seen from (5.29) that in order for a solution to exist, the functions $p(t)$ and $q(t)$ must meet the following ODEs

[397] See Equations (3.7),(3.36) and (3.37).

$$(\gamma - \kappa_r p(t) + p'(t)) = 0 \tag{5.30}$$

and

$$-e^{(t-T)\kappa_r}\gamma + \gamma + \kappa_r\left(q'(t) - \kappa_\varepsilon q(t)\right) = 0 \tag{5.31}$$

We first solve the ODE for the unknown function p. Its solution is

$$p(t) = \frac{\left(1 - e^{-(T-t)\kappa_r}\right)\gamma}{\kappa_r} \tag{5.32}$$

We now insert the solution for $p(t)$ into the second ODE for $q(t)$ and obtain[398]

$$q(t) = \frac{\gamma\left(\left(1 - e^{(t-T)\kappa_\varepsilon}\right)\kappa_r + \left(-1 + e^{(t-T)\kappa_r}\right)\kappa_\varepsilon\right)}{\kappa_r\left(\kappa_r - \kappa_\varepsilon\right)\kappa_\varepsilon} \tag{5.33}$$

We notice from (5.32), (5.32), (3.45) and (3.46) that the following relationships hold

$$p(t) = \gamma B_1(t,T) \tag{5.34}$$

$$q(t) = \gamma B_2(t,T) \tag{5.35}$$

We insert $p(t)$ and $q(t)$ into (5.29) and note that it becomes an ODE for the unknown function $g(t)$. The solution to the function $g(t)$ is once again not needed for the computation of the optimum portfolio weights, hence we give its solution only in the Appendix.[399] $p(t)$ from (5.33), $q(t)$ from (5.33) and $g(t)$ from (C.2) hence solve Equation (5.29). Now, we are able to give the solution to the optimum zero-coupon bond portfolio weights. We insert $p(t)$ from (5.34) and $q(t)$ from (5.35) into Equation (5.22). We obtain

$$w^* = \frac{1}{1-\gamma}\left(\Sigma^{-1}\sigma\lambda\right) + \frac{\gamma}{\gamma-1}\left(\Sigma^{-1}\hat{\Sigma}\begin{pmatrix}B_1(t,T)\\B_2(t,T)\end{pmatrix}\right)$$

or

$$w^* = \frac{1}{1-\gamma}\left(\Sigma^{-1}\sigma\lambda\right) - \sum_{i=1}^{2}\frac{\gamma}{1-\gamma}B_i(t,T)\Sigma^{-1}\hat{\Sigma}_i \tag{5.36}$$

where $\hat{\Sigma}_i$ is again the i-th column of the matrix $\hat{\Sigma} \equiv \sigma\beta'$. The optimum portfolio weights are again an affine combination of three portfolios – the mean-variance efficient portfolio and two hedge portfolios.

Interpretation of Optimum Portfolio Weights

Since the structure of the mean-variance efficient portfolio is the same as in the Vasicek case, we concentrate on analyzing the hedge portfolios. The hedge

398 We can find a solution for $g(t)$ too, but since this function is not necessary for the calculation of the optimum portfolio weights we refrain from giving it here.

399 See Appendix C.2, Equation (C.2).

portfolio against the factor i ($f_1 = r$ and $f_2 = \varepsilon$) can be written in detail as follows

$$-\frac{\gamma}{1-\gamma}B_i(t,T)\begin{pmatrix}\frac{\text{var}\left(\frac{dP_2}{P_2}\right)\text{cov}\left(\frac{dP_1}{P_1},df_i\right)-\text{cov}\left(\frac{dP_1}{P_1},\frac{dP_2}{P_2}\right)\text{cov}\left(\frac{dP_2}{P_2},df_i\right)}{\text{var}\left(\frac{dP_1}{P_1}\right)\text{var}\left(\frac{dP_2}{P_2}\right)-\text{cov}\left(\frac{dP_1}{P_1},\frac{dP_2}{P_2}\right)^2}\\ \frac{\text{var}\left(\frac{dP_1}{P_1}\right)\text{cov}\left(\frac{dP_2}{P_2},df_i\right)-\text{cov}\left(\frac{dP_1}{P_1},\frac{dP_2}{P_2}\right)\text{cov}\left(\frac{dP_1}{P_1},df_i\right)}{\text{var}\left(\frac{dP_1}{P_1}\right)\text{var}\left(\frac{dP_2}{P_2}\right)-\text{cov}\left(\frac{dP_1}{P_1},\frac{dP_2}{P_2}\right)^2}\end{pmatrix} \tag{5.37}$$

with the approaching of the investment horizon T, the hedging volume[400] becomes smaller since $\frac{\partial B_i(t,T)}{\partial t} < 0$ and it vanishes when $t = T$. It is interesting to note from Equation (5.37) that the correlation between the factors r and ε seems to play no explicit role. Nevertheless ϱ has seemingly an influence on the various covariances. But if we calculate the different covariances and insert them into (5.37), the expression reduces to

$$-\frac{\gamma}{1-\gamma}B_1(t,T)\begin{pmatrix}\frac{B_2(t,T_2)}{B_1(t,T_1)B_2(t,T_2)-B_1(t,T_2)B_2(t,T_1)}\\ \frac{B_2(t,T_1)}{B_1(t,T_2)B_2(t,T_1)-B_1(t,T_1)B_2(t,T_2)}\end{pmatrix} \tag{5.38}$$

for the first factor (r) and to

$$-\frac{1}{1-\gamma}B_2(t,T)\begin{pmatrix}\frac{B_1(t,T_2)}{B_1(t,T_2)B_2(t,T_1)-B_1(t,T_1)B_2(t,T_2)}\\ \frac{B_1(t,T_1)}{B_1(t,T_1)B_2(t,T_2)-B_1(t,T_2)B_2(t,T_1)}\end{pmatrix} \tag{5.39}$$

for the second factor (ε), with $B_1(t,T)$ and $B_2(t,T)$ defined in equations (3.45) and (3.46) respectively. The hedge portfolios therefore don't depend on the correlation coefficient between the two factors. Mathematically this follows since the covariances are all linear in ϱ and therefore we can reduce ϱ accordingly. Economically this is due to the fact that the term structure model is affine linear and that all zero-coupon bonds are influenced by both factors, i.e. there is no idiosyncratic risk.

The hedge portfolio volumes are always non-negative since $B_1(t,T) \geq 0$ and $B_2(t,T) \geq 0$.[401] The first inequality is easy to show, the second needs more attention. To see this, note that $B_2(t,T)$ can only be negative if (i) the nominator is positive and the denominator is negative or (ii) the nominator is negative and the denominator is positive. The sign of the denominator depends only on the relationship between κ_r and κ_ε, if $\kappa_r > \kappa_\varepsilon$ then the denominator is positive and vice versa. The sign of the nominator depends on a more complicated expression. If

$$\frac{1-e^{-(T-t)\kappa_\varepsilon}}{1-e^{-(T-t)\kappa_r}} > \frac{\kappa_\varepsilon}{\kappa_r}$$

then the nominator is positive and vice versa. It can then be shown that the above mentioned constellations for a negative function $B_2(t,T)$ cannot occur for permissible values of the parameters.

400 $-\frac{\gamma}{1-\gamma}B_i(t,T)$

401 And $0 < \gamma < 1$.

Another interesting finding is that the signs of the portfolio weights for each hedge portfolio are different, i.e. if the hedge portfolio for factor f_i contains a long (short) position in the T_1-zero-coupon bond, then it also contains a short (long) position in the T_2-zero-coupon bonds. To see this, note that the following relationship holds for denominators in the asset weights formula

$$B_1(t,T_1)B_2(t,T_2) - B_1(t,T_2)B_2(t,T_1) = \\ - [B_1(t,T_2)B_2(t,T_1) - B_1(t,T_1)B_2(t,T_2)]$$

Comparing the zero-coupon bond weights from Equations (5.38) and (5.39), we also note that if there is a long (short) position in the τ-zero-coupon bond in the r-hedge portfolio, there must be a short (long) position in this bond in the ε-hedge portfolio.[402]

Numerical Example

We want to conclude with a numerical example. We assume the same parameters for the HW2 model as in the corresponding section in Chapter 4.2.5 on Markowitz Portfolio Selection, i.e. $r(0) = 0.025$, $\varrho = 0.6$, $\lambda_1 = 1.2395$, $\lambda_2 = 0$, $\sigma_r = 0.0073$, $\sigma_\varepsilon = 0.0219$, $\kappa_r = 0.2591$, $\kappa_\varepsilon = 0.8274$ and $\theta = 0.0053$.

For this parametrization with an investment horizon of $T = 5$, two risky zero-coupon bonds with maturity $T_1 = 10$ and $T_2 = 30$ and a risk aversion parameter $\gamma = 0.5$, we obtain the following portfolio weights at time 0:

$$w^* = \begin{pmatrix} 2593.44 \\ -2312.79 \end{pmatrix} \tag{5.40}$$

This result can be disaggregated as follows:

Table 5.3. HW2 model: Disaggregation of optimum portfolio.

Asset	Mean-Variance	Hedge Portfolio 1	Hedge Portfolio 2	Sum
10-year zero	2,596.91	-21.43	17.97	2,593.44
30-year zero	-2,315.27	19.11	-16.63	-2,312.79
MMA	—	—	—	-279.65

It is highly unlikely that such a portfolio would or could be implemented in practice since it requires huge long and short positions. In this respect, the result is similar to the unrestricted static model. The reason is the extreme dependence on the market prices of interest rate risk, λ_1 and λ_2. Even small changes in λ_1 or λ_2 have a pronounced effect on the portfolio weights.[403]

[402] Compare again the denominators and use the fact that $B_1(t,T)$ and $B_2(t,T)$ are non-negative.

[403] The table gives the portfolio vector (w_1, w_2, w_0) where w_1 (w_2) is the weight of the 10- year zero-coupon bond (30-year zero-coupon bond) and w_0 is the weight of the money market account.

Table 5.4. HW2 model: Effect of market prices of interest rate risk on portfolio weights.

λ_2 / λ_1	0	0.05	0.1	0.15	0.2
0	$\begin{pmatrix}-3.47\\2.48\\1.98\end{pmatrix}$	$\begin{pmatrix}-118.15\\105.96\\13.2\end{pmatrix}$	$\begin{pmatrix}-232.84\\209.43\\24.41\end{pmatrix}$	$\begin{pmatrix}-347.53\\312.91\\35.63\end{pmatrix}$	$\begin{pmatrix}-462.22\\416.38\\46.84\end{pmatrix}$
0.05	$\begin{pmatrix}101.29\\-90.91\\-9.38\end{pmatrix}$	$\begin{pmatrix}-13.4\\12.56\\1.84\end{pmatrix}$	$\begin{pmatrix}-128.09\\116.04\\13.05\end{pmatrix}$	$\begin{pmatrix}-242.78\\219.51\\24.27\end{pmatrix}$	$\begin{pmatrix}-357.46\\322.99\\35.48\end{pmatrix}$
0.1	$\begin{pmatrix}206.05\\-184.31\\-20.74\end{pmatrix}$	$\begin{pmatrix}91.36\\-80.84\\-9.52\end{pmatrix}$	$\begin{pmatrix}-23.33\\22.64\\1.69\end{pmatrix}$	$\begin{pmatrix}-138.02\\126.12\\12.9\end{pmatrix}$	$\begin{pmatrix}-252.71\\229.59\\24.12\end{pmatrix}$
0.15	$\begin{pmatrix}310.8\\-277.71\\-32.1\end{pmatrix}$	$\begin{pmatrix}196.11\\-174.23\\-20.88\end{pmatrix}$	$\begin{pmatrix}81.43\\-70.76\\-9.67\end{pmatrix}$	$\begin{pmatrix}-33.26\\32.72\\1.54\end{pmatrix}$	$\begin{pmatrix}-147.95\\136.19\\12.76\end{pmatrix}$
0.2	$\begin{pmatrix}415.56\\-371.1\\-43.46\end{pmatrix}$	$\begin{pmatrix}300.87\\-267.63\\-32.25\end{pmatrix}$	$\begin{pmatrix}186.18\\-164.15\\-21.03\end{pmatrix}$	$\begin{pmatrix}71.49\\-60.68\\-9.82\end{pmatrix}$	$\begin{pmatrix}-43.2\\42.8\\1.4\end{pmatrix}$

This table summarizes the portfolio weight sensitivities to changes in market prices of interest rate risk. In this numerical example, even small changes in the market prices have a huge impact on the portfolio weights. But we observe more structure. For a given λ_2, a higher value of λ_1 leads to lower weights for the shorter bond and higher weights for the longer bond. For a given λ_1, higher values of λ_2 lead to higher weights for the shorter bond and lower weights for the longer bond. No values of λ_1 and λ_2 produce sensible portfolio weights ($0 \leq w_i \leq 1$) in this example.

The obvious solution of introducing short sale constraints into the optimization problem unfortunately makes the problem much more difficult so solve. Constrained optimization problems in continuous-time have been studied by Cvitanic/Karatzas (1992), Xu/Shreve (1992a) and Xu/Shreve (1992b). We will not solve this constrained optimization problem here.

5.4 International Bond Investing

5.4.1 Introduction

The general dynamic bond portfolio selection problem presented in the last sections can be easily adapted to other settings. A theoretically interesting and practically important extension is to consider foreign investments, i.e. the inclusion of foreign currency bonds.[404] Frequently, the investors don't have regional investment restrictions and hence can invest not only in EUR-denominated Government bonds but also in Government bonds denominated

[404] The general foreign currency framework is due to Amin/Jarrow (1991).

in a foreign currency, e.g. US Treasury Bonds or Japanese Government Bonds. Modern portfolio theory propagates the inclusion of non-perfectly correlated assets in the investment universe because this generates diversification benefits. Foreign currency Government bonds are then a natural choice for the investor.

One major difficulty of portfolio optimization problems in an international setting is modeling all asset prices in an arbitrage-free manner. In this section we base the setup of the foreign exchange market on Lipton (2001). He used comparable frameworks to study the valuation of foreign exchange derivatives. We use his modeling approach to study an international bond portfolio selection model.[405]

5.4.2 Model Setup

There exist many possibilities of modeling international bond markets. One can decide on the number of countries one wishes to model, the term structure of interest rate model in each country (one-factor, multi-factor) and the stochastic process for the foreign exchange rate. In this section we restrict our attention to a simple two-country model where the term structure in each country is governed by the Vasicek term structure model.

- The investor can invest his funds in different assets in two countries[406]
  - Country 1 (Germany, currency EUR) and
  - Country 2 (USA, currency USD)
- In every country the term structure of interest rates is determined by the Vasicek (1977) term structure model with constant market prices of interest rate risk as introduced in Chapter 3.4. The short rate dynamics are
  - Country 1
  $$dr_1(t) = \kappa_1(\theta_1 - r_1)dt + \sigma_{r,1}dz_1 \tag{5.41}$$
  - Country 2
  $$dr_2(t) = \kappa_2(\theta_2 - r_2)dt + \sigma_{r,2}dz_2 \tag{5.42}$$

  We assume that dz_1 and dz_2 are uncorrelated.
- The EUR/USD exchange rate $S(t)$ is governed by a two-dimensional Itô process. The exchange rate follows
  $$\frac{dS(t)}{S(t)} = a(.)dt + b_1(.)dz_1 + b_2(.)dz_2 \tag{5.43}$$
  where a is the drift and b_1 and b_2 are the volatilities of the exchange rate.

---

[405] To our knowledge this international bond portfolio selection model presented here, constitutes an addition to the existing literature.

[406] For ease of exposition these countries are taken to be Germany and the United States of America.

First, we examine Germany (country 1). In Germany, a money market account trades. The dynamics of the money market account is given by the following SDE

$$\frac{dM_1(t)}{M_1(t)} = r_1(t)dt$$

where $r_1(t)$ is the (nominal) short rate of interest. Also zero-coupon bonds of different maturities trade with SDEs

$$\frac{dP_1(t,T)}{P_1(t,T)} = (r_1(t) + \sigma_{1,1}(t,T)\lambda_1(t))dt - \sigma_{1,1}(t,T)dz_1$$

where $\lambda_1(t)$ is the z_1-market price of risk in country 1 and $\sigma_{1,1}(t,T)$ is the volatility of the zero-coupon bond prices. Next, we consider the USA (country 2). In the USA, there is also a market for a money market account. It follows the SDE

$$\frac{dM_2(t)}{M_2(t)} = r_2(t)dt$$

where $r_2(t)$ is the short rate of interest. There also trade zero-coupon bonds of different maturities. The SDEs for the bonds are

$$\frac{dP_2(t,T)}{P_2(t,T)} = (r_2(t) + \sigma_{2,2}(t,T)\lambda_2^*(t))dt - \sigma_{2,2}(t,T)dz_2$$

where $\lambda_2^*(t)$ is the z_2-market price of interest rate risk in USA and $\sigma_{2,2}(t,T)$ is the volatility of the zero-coupon bond prices.

We will see that since the currencies are freely convertible, the drift and the volatilities of the exchange rate can't be set independently of the bond markets. We now construct an arbitrage free international bond market, i.e. we find that drift a of the exchange rate that guarantees freedom of arbitrage. Our derivation follows Lipton (2001).[407]

First, we convert USD-prices to EUR

$$M^*(t) = M_2(t)S(t)$$
$$P^*(t,T) = P_2(t,T)S(t)$$

The dynamics of both converted prices can be obtained by a straightforward application of Itô's lemma. The MMA dynamics are

$$\frac{dM^*(t)}{M^*(t)} = (r_2 + a)dt + b_1dz_1 + b_2dz_2 \tag{5.44}$$

The dynamics of the USD zero-coupon bond in EUR are obtained in a similar way

[407] See Lipton (2001), pp. 223–227.

$$\begin{aligned}
&\frac{dP^*(t,T)}{P^*(t,T)} \\
&= (r_2(t) + a + \sigma_{2,2}(t,T)(\lambda_2^*(t) - b_2))dt + b_1 dz_1 + (b_2 - \sigma_{2,2}(t,T))dz_2 \\
&= (r_2(t) + a + \sigma_{2,2}(t,T)\lambda_2(t))dt + b_1 dz_1 + (b_2 - \sigma_{2,2}(t,T))dz_2 \qquad (5.45)
\end{aligned}$$

where we have defined $\lambda_2(t) := \lambda_2^*(t) - b_2$. We now construct a riskless portfolio from holdings in P^*, M^* and P_1. The instantaneous return of this riskless portfolio must be equal to the short rate in Germany (country 1), i.e. r_1. We solve for a and obtain

$$a = r_1 - r_2 - \lambda_1 b_1 - \lambda_2 b_2 \qquad (5.46)$$

We can now insert a in the equations for S, P^* and M^* but we can gain further insights if we examine $1/S$ first. The dynamics of $S^* = 1/S$ can again be obtained in a straightforward manner by Itô's lemma

$$dS^* = a^* dt + b_1^* dz_1 + b_2^* dz_2$$

where

$$\begin{aligned}
a^* &= -a + b_1^2 + b_2^2 \\
b_1^* &= -b_1 \\
b_2^* &= -b_2
\end{aligned}$$

Arbitrage-free considerations[408] lead to the following functional form for a^*

$$\begin{aligned}
a^* &= r_2 - r_1 - (\lambda_1 - b_1^*)b_1^* - \lambda_2^* b_2^* \\
&= r_2 - r_1 - (\lambda_1 + b_1)b_1^* - \lambda_2^* b_2^* \\
&= r_2 - r_2 - \lambda_1^* b_1^* - \lambda_2^* b_2^*
\end{aligned}$$

where we defined $\lambda_1^* := \lambda_1 + b_1$. We now obtain the following relationships

$$b_1 = \lambda_1^* - \lambda_1 \qquad (5.47)$$
$$b_2 = \lambda_2^* - \lambda_2 \qquad (5.48)$$

The foreign exchange volatilities are determined by the excess rates of return on discount bonds and cannot be chosen independently of the bond markets under consideration.[409] Equation (5.43) can hence be written as

$$\frac{dS(t)}{S(t)} = (r_1 - r_2 - \lambda_1(\lambda_1^* - \lambda) - \lambda_2(\lambda_2^* - \lambda_2))dt + (\lambda_1^* - \lambda_1)dz_1 + (\lambda_2^* - \lambda_2)dz_2$$

The drift of the exchange rate is completely determined by the interest rate differential and the market prices of interest rate risk in the two countries.

[408] Construction of a riskless portfolio.

[409] See Flesaker/Hughston (2000), p. 220.

The price dynamics for the USD money market account and the USD zero-coupon bond in EUR from (5.44) and (5.45) can now be rewritten with a from (5.46) and b_1 and b_2 from (5.47) and (5.48) as follows

$$\begin{aligned}\frac{dM^*(t)}{M^*(t)} =&(r_1(t) + (\lambda_1 - \lambda_1^*)\lambda_1 + (\lambda_2 - \lambda_2^*)\lambda_2)dt \\ &- (\lambda_1 - \lambda_2)dz_1 - (\lambda_2 - \lambda_2^*)dz_2 \end{aligned} \tag{5.49}$$

and

$$\begin{aligned}\frac{dP^*(t,T)}{P^*(t,T)} =&(r_1(t) + (\lambda_1 - \lambda_1^*)\lambda_1 + (\lambda_2 - \lambda_2^* + \sigma_{2,2}(t,T))\lambda_2)dt \\ &- (\lambda_1 - \lambda_1^*)dz_1 - (\lambda_2 - \lambda_2^* + \sigma_{2,2}(t,T))dz_2 \end{aligned} \tag{5.50}$$

In this simple example, the instantaneous return of the foreign currency assets in local currency is completely determined by the local short rate r_1 since we assumed constant market prices of interest rate risk.

5.4.3 Derivation of the Optimum Portfolio Weights

The investor can invest his initial wealth in a local zero-coupon bond of maturity T_1, a foreign zero-coupon bond of maturity T_2 and a riskless money market account that yields the local riskfree rate.

We now solve the international bond portfolio selection problem in accordance with the method outlined at the beginning of the chapter. We use Equation (5.2) with[410]

$$\sigma_t = \begin{pmatrix} \sigma_{1,1}(t,T_1) & 0 \\ \lambda_1 - \lambda_1^* & \sigma_{2,2}(t,T_2) + \lambda_2 - \lambda_2^* \end{pmatrix} \tag{5.51}$$

and obtain the following expression for the wealth dynamics

$$\frac{dW}{W} = \mu_W(r_1,t,T_1,T_2)dt - \sigma_{W,1}(t)dz_1 - \sigma_{W,2}(t)dz_2 \tag{5.52}$$

where

$$\begin{aligned}\mu_W(r_1,t,T_1,T_2) =&r_1 + w_1\sigma_{1,1}(t,T_1)\lambda_1 + w_2((\lambda_1 - \lambda_1^*)\lambda_1 \\ &+ (\lambda_2 - \lambda_2^* + \sigma_{2,2}(t,T_2))\lambda_2) \\ \sigma_{W,1}(t) =&w_1\sigma_{1,1}(t,T_1) + w_2(\lambda_1 - \lambda_1^*) \\ \sigma_{W,2}(t) =&w_2(\lambda_2 - \lambda_2^* + \sigma_{2,2}(t,T_2))\end{aligned}$$

Because of arbitrage-relationships and the characteristics of the Vasicek term structure model (here: assumption of constant market price of interest rate

[410] The German zero-coupon bond has maturity date T_1 and the US zero-coupon bond T_2.

risk), the wealth dynamics depend on only two factors, namely W and r_1. The exchange rate and the short rate in the USA r_2 don't influence the wealth dynamics of the investor.

We solve the problem analogous to the Vasicek case in section 5.3.1. The uncertainty is driven by one state variable r_1 only – hence $x(t) = r_1(t)$ – and two uncorrelated Brownian motions $z_1(t)$ and $z_2(t)$. We hence need two positions in risky assets and a position in the riskless asset, i.e. $k = 1$, $d = n = 2$. Since $k = 1$, our optimum portfolio will consist of a mean-variance efficient portfolio and a single hedge portfolio.

We can specify the general portfolio weight equation for a CRRA investor in Equation (5.22) as follows

$$\sigma = \begin{pmatrix} \sigma_{1,1}(t,T_1) & 0 \\ \lambda_1 - \lambda_1^* & \lambda_2 - \lambda_2^* + \sigma_{2,2}(t,T_2) \end{pmatrix}$$

$$\lambda = \begin{pmatrix} \lambda_1 \\ \lambda_2 \end{pmatrix}$$

$$\beta = \begin{pmatrix} \sigma_{r,1} & 0 \end{pmatrix}$$

We assume the following functional form for the separation function $f(r_1, t)$. This is the same assumption as in the section on the Vasicek model.[411]

$$f(r_1, t) = g(t) \exp(v(t) r_1)$$

then

$$\frac{f_{r_1}}{f} = v(t)$$

The zero-coupon bond volatilities are[412]

$$\sigma_{1,1}(t,T) = \sigma_{r_1} B_1(t,T) = \sigma_{r_1} \left(\frac{1 - e^{-\kappa_1 (T-t)}}{\kappa_1} \right) \tag{5.53}$$

$$\sigma_{2,2}(t,T) = \sigma_{r_2} B_2(t,T) = \sigma_{r_2} \left(\frac{1 - e^{-\kappa_2 (T-t)}}{\kappa_2} \right) \tag{5.54}$$

Inserting the assumption for f in Equation (5.21) leads to

$$\begin{aligned} e^{r_1 v(t)} r_1 g(t) \left(\gamma - \kappa_1 v(t) + v'(t)\right) + \frac{e^{r_1 v(t)}}{2(\gamma - 1)} \\ \times \left(2(\gamma - 1) g'(t) - g(t) \left(\gamma \lambda_1^2 - 2\gamma \sigma_{r_1} v(t) \lambda_1 + \gamma \lambda_2^2 \right.\right. \\ \left.\left. + v(t) \left(\sigma_{r_1}^2 v(t) - 2(\gamma - 1)\theta_1 \kappa_1\right)\right)\right) = 0 \end{aligned} \tag{5.55}$$

[411] See Equation 5.24.

[412] Each zero-coupon bond price is given by the Vasicek bond price formula and so the volatilities are the same as in Chapter 5.3.1.

The expression $\gamma - \kappa_1 v(t) + v'(t)$ must hence be zero for all values of r_1. The solution to $v(t)$ is the same as in the Vasicek case, i.e.

$$v(t) = \gamma \frac{1 - \exp(-\kappa_1(T-t))}{\kappa_1} = \gamma B_1(t,T) \tag{5.56}$$

With $v(t)$ from (5.56) Equation (5.55) becomes an ODE for the unknown function $g(t)$. The solution for $g(t)$ is once again not needed for the derivation of the optimum portfolio weight, we give it therefore only in the Appendix.[413] $v(t)$ and $g(t)$ solve Equation (5.55). With the solution for $v(t)$ we obtain the optimum portfolio weights

$$w^* = \frac{1}{1-\gamma}\left(\Sigma^{-1}\sigma\lambda\right) - \frac{\gamma}{1-\gamma}\Sigma^{-1}\hat{\Sigma}B_1(t,T) \tag{5.57}$$

The hedge term can be simplified further and we eventually obtain

$$w^* = \frac{1}{1-\gamma}\left(\Sigma^{-1}\sigma\lambda\right) - \frac{\gamma}{1-\gamma}B_1(t,T)\begin{pmatrix}\frac{1}{B_1(t,T_1)}\\ 0\end{pmatrix} \tag{5.58}$$

5.4.4 Interpretation of the Optimum Portfolio Weights

We now have a closer look at the optimum portfolio weights from an economic point of view. A general discussion of the influence of the parameters on the mean-variance portfolio is difficult. The mean-variance efficient portfolio can be written as

$$\begin{pmatrix}\frac{\lambda_2\lambda_1^* + \lambda_1(\sigma_2(t,T_2) - \lambda_2^*)}{\sigma_1(t,T_1)\left(\lambda_2 - \lambda_2^* + \sigma_2(t,T_2)\right)}\\ \frac{\lambda_2}{\lambda_2 - \lambda_2^* + \sigma_2(t,T_2)}\end{pmatrix} \tag{5.59}$$

Utilizing the relationships from (5.47) and (5.48) we can write this as

$$\begin{pmatrix}\frac{b_1\lambda_2 + (\sigma_2(t,T_2) - b_2)\lambda_1}{\sigma_1(t,T_1)(\sigma_2(t,T_2) - b_2)}\\ \frac{\lambda_2}{\sigma_2(t,T_2) - b_2}\end{pmatrix} \tag{5.60}$$

If we assume positive market prices of interest rate risk and positive volatilities of the exchange rate, then both portfolio weights are positive if $\sigma_2(t,T_2) > b_2$.

The hedge portfolio in Equation (5.58) consists of a position in the local zero-coupon bond only. This seems counterintuitive at first, but is perfectly rational. We have stated before[414], that the hedge portfolio is formed in such a way as to provide the best possible hedge against the state variables (here r_1), in this case this is a position only in the local zero-coupon bond since the instantaneous correlation coefficient between dr_1 and $\frac{dP_1(t,T)}{P_1(t,T)}$ is -1. An additional position in the foreign zero-coupon bond would reduce the (absolute) correlation and therefore diminish the effect of the hedge portfolio.

[413] See Appendix C.3 , Equation (C.3).

[414] See Section 5.2.3 on page 93.

It is interesting to note that we can recover the single currency Vasicek (1977) problem[415] as a special case by setting $\lambda_2 = 0$. Then the weight of the second bond in the mean-variance efficient portfolio becomes zero, i.e. a position in the second zero-coupon bond is only taken when λ_2 is not equal to zero.

5.4.5 Numerical Example

We now demonstrate the applicability of the model by means of a numerical example. We assume the following values for the parameters of the term structure models, the zero-coupon bonds, the foreign exchange rate volatilities and the investor's risk aversion and investment horizon:

- Investor
  - Risk aversion parameter: $\gamma = 0.5$
  - Investment horizon: $T = 2$
- Term structure
  - Germany: $\kappa_1 = 0.7$, $\sigma_{r_1} = 0.02$, $\lambda_1 = \lambda_2 = 0.01$
  - USA: $\kappa_2 = 0.8$, $\sigma_{r_2} = 0.03$
- Zero-coupon bond maturities: $T_1 = T_2 = 10$
- Exchange rate: $b_1 = b_2 = 0.01$

We insert the parameters into Equation (5.58) and obtain

$$w^* = \begin{pmatrix} 0.0878182 \\ 0.403261 \end{pmatrix} \tag{5.61}$$

The resulting portfolio has long positions in both zero-coupon bonds and in the riskless money market account. This result can be disaggregated as follows

Table 5.5. International bond portfolio selection: Disaggregation of optimum portfolio weights.

Weight	Mean-variance	Hedge portfolio	Overall
Local zero	0.841909	-0.754091	0.0878182
Foreign zero	0.403261	0	0.403261
Local MMA	—	—	0.508921

These results seem quite plausible but as has been said in the last two sections where we covered the Vasicek and the HW2 models, the optimum portfolio weights are heavily dependent on the parameters, especially the unobservable market prices of interest rate risk.

[415] The solution was given in Chapter 5.3.1.

5.5 Summary and Conclusion

The dynamic portfolio selection framework is quite flexible and analytical solutions can be derived for simple interest rate models. We derived the optimum portfolio strategy for CRRA investors in the Vasicek and HW2 case. The results for the Vasicek model were already published in Korn/Kraft (2002) but the specific results for the HW2 model were new.[416] We have shown that the framework can be extended to foreign exchange risks. In the numerical examples the resulting portfolio weights unfortunately displayed huge long and short positions in the assets. It is highly unlikely that these extreme positions (in terms of position size) would ever be implemented in practice.

The results mirror the outcomes of the unrestricted static portfolio selection in Chapter 4. There, we could improve the plausibility of the portfolios by introducing short-sale constraints. Constrained optimization problems in continuous-time have been studied by Cvitanic/Karatzas (1992), Xu/Shreve (1992a) and Xu/Shreve (1992b). For more complicated (multi-factor) models or constrained optimization problems, numerical methods must be used.[417]

---

[416] Munk/Sørensen (2004) solve the dynamic optimization problem for HJM term structure movements. The HW2 model is a special case of the HJM framework, so the optimum portfolio weights could have been recovered from them.

[417] For a textbook treatment see Kushner/Dupuis (2000).

6 Summary and Conclusion

The aim of this thesis was to analyze the necessary adjustments to static and dynamic models of equity portfolio optimization in order to use these models for the selection of bond portfolios. Furthermore, we wanted to compare the optimum portfolios with bond portfolios recommended by the practice. The core of this thesis is based mainly upon Wilhelm (1992) and Korn/Kraft (2002).

We derived the general HJM framework for term structure modeling in Chapter 3. The derivation employed the stochastic discount factor approach and not the normally utilized martingale or PDE approaches. Then we obtained the Vasicek (1977) and the HW2 models as special cases of the general HJM framework.

Chapter 4 dealt with static bond portfolio optimization methods. We presented the bond portfolio selection model by Wilhelm (1992) and showed in detail how the necessary parameters could be derived. Wilhelm (1992) showed how to use this model in conjunction with the CIR term structure model. We extended his approach to the Vasicek and the HW2 models. We found that when the number of assets is far greater than the number of risk factors, the mean-variance efficient portfolios contained huge long and short positions. This finding hence confirms the conclusion reached by Korn/Koziol (2006). It is very unlikely that these portfolios could be implemented in practice. Korn/Koziol (2006) suggest limiting the number of bonds to include in the portfolio. We used short-sale constrained optimization to find optimum bond portfolios in this thesis. A drawback of this approach is that no analytic solutions can be found for the bond portfolio selection problem, so we had to confine our analysis to numerical examples. The resulting portfolios contained oftentimes only positions in relatively few zero-coupon bonds. The comparison of mean-variance efficient portfolios with portfolio propagated by the practice produced interesting results. Portfolios resulting from the duration and yield curve strategies could not in all cases be reproduced with the mean-variance framework. It seemed that particular portfolios of bonds are preferred regardless of the expectations about future term structure movements. Obviously, the main driver for these differences, is taking into consideration the risks of these strategies in the mean-variance framework. On the other hand, the duration-immunized portfolios performed quite well in the Vasicek and the HW2 term structure models. Hence, the simple portfolio rule for implementing these portfolios can be seen as a reasonable approximation to reality.

In Chapter 5, we presented dynamic models of bond portfolio selection. These are based on the seminal paper by Merton (1969). His approach can be generalized to interest rate sensitive assets as well. We presented the optimum portfolio strategy for CRRA investors in a Vasicek model. This result was obtained previously by Korn/Kraft (2002). Then we showed how to obtain an explicit solution for a bond portfolio selection problem with the HW2 model using the traditional stochastic control approach. This result presents an addition to the bond portfolio literature.[418] We furthermore analyzed a simple extension of this framework, an international bond portfolio selection problem. We were able to derive an analytic expression for the optimum portfolio strategy in a two country case. Unfortunately, the unconstrained portfolios generally contained large long and short positions. This finding is similar to the results we obtained from static portfolio optimization. Constraining short sales seems to be the obvious measure, but then the resulting optimization problem is far more complicated. We refrained from solving constrained dynamic optimization problems in this thesis.

An application of the models in practice is in our opinion difficult. At the moment, I think, virtually no mutual fund company employs sophisticated dynamic optimization models for (bond) portfolio selection. Hence, an implementation of the static mean-variance model for bond portfolios seems more probable. There is a good chance that quantitatively oriented bond portfolio managers give this approach a try. But before even those portfolio managers consider an application of the model, a backtesting has to be performed.[419]

For an implementation, we recommend a two-factor term structure model, e.g. the HW2 model. Principal component analysis of the term structure showed that two to three factors account for the vast majority of the variation in the term structure.[420] A two-factor model produces also realistic correlations between bonds of different maturities.[421] A problem that might arise is the limited number of bonds the mean-variance approach typically advises to buy. Hence, the typical government bond portfolio would consist of only a handful of assets. This concentration of the portfolio value in just a couple of assets might make the client nervous and furthermore (if the fund has significant money to invest) might affect the prices in the bond market. This is usually refered to as the large-trader problem.[422] Another obstacle is

---

418 Munk/Sørensen (2004) derived optimum bond portfolios for HJM models using the martingale approach.

419 This has already been examined for the German market by Korn/Koziol (2006).

420 Litterman/Scheinkman (1991).

421 In a one-factor model, all bond prices are perfectly correlated.

the prevailing relative portfolio management approach, i.e. the management of a portfolio relative to a benchmark. Mean-variance efficient portfolios may deviate severely from the composition (and possibly from the risk characteristics) of the benchmark, hence these portfolios will probably have a large tracking error.[423]

Despite these difficulties, we think this approach is promising. It is based on an accepted theory and is widely employed in equity portfolio selection. In our opinion it is better suited to cope with the complex fixed income risks than the ad hoc approaches currently employed in practice.

[422] For consumption and investment problems with a large trader, see Cuoco/Cvitanic (1998) and Bank/Baum (2004).

[423] The investment guidelines might limit the tracking error of the portfolio.

A

Heath/Jarrow/Morton (1992)

A.1 Dynamics of Zero-Coupon Bonds

The zero-coupon bond price can be written as a function of the instantaneous forward rate curve as follows[1]

$$P(t,T) = \exp\left(-\int_t^T f(t,u)\,du\right) \tag{A.1}$$

Let's define $Y(t) = \int_t^T f(t,u)\,du$. The problem is now to determine the dynamics of $Y(t)$ given the dynamics of the forward rate curve in (3.2). It has been shown by Heath/Jarrow/Morton (1992) by using a generalized stochastic Fubini theorem, that the dynamics of Y can be written as[2]

$$dY(t) = \left[\left(\int_t^T m(t,u)\,du\right) - f(t,t)\right] dt + \sum_{i=1}^{d}\left(\int_t^T s_i(t,u)\,du\right) dz_i(t) \tag{A.2}$$

Since $P(t,T) = g(Y(t))$ with $g(Y) = \exp(-Y)$, $g'(Y) = -\exp(-Y)$ and $g''(Y) = \exp(-Y)$ it follows by application of Itô's lemma that the dynamics of $P(t,T)$ are

$$\begin{aligned} dP(t,T) &= g'(Y)dY + \frac{1}{2}g''(Y)(dY)^2 \\ &= -\exp(-Y)dY + \frac{1}{2}\exp(-Y)(dY)^2 \\ &= -P(t,T)dY + \frac{1}{2}P(t,T)(dY)^2 \end{aligned}$$

Inserting dY and recognizing that

[1] See Equation (2.10).

[2] See Heath/Jarrow/Morton (1992), p. 99.

$$(dY)^2 = \sum_{i=1}^{d} \left(\int_t^T s_i(t,u)\, du \right)^2 dt$$

we obtain the following formula for the evolution of $P(t,T)$

$$\frac{dP(t,T)}{P(t,T)} = \left[f(t,t) - \left(\int_t^T m(t,u)\, du \right) + \frac{1}{2} \sum_{i=1}^{d} \left(\int_t^T s_i(t,u)\, du \right)^2 \right] dt$$
$$- \sum_{i=1}^{d} \left(\int_t^T s_i(t,u)\, du \right) dz_i(t) dt \tag{A.3}$$

A.2 Arbitrage-Free Pricing

Assume that $x = \ln(\zeta)$, then it follows from Equation (3.11) and an application of Itô's lemma that

$$dx = x_t dt + x_\zeta d\zeta + \frac{1}{2} x_{\zeta\zeta} (d\zeta)^2$$

with

$$\begin{aligned} x_t &= \frac{\partial x}{\partial t} = 0 \\ x_\zeta &= \frac{\partial x}{\partial \zeta} = \frac{1}{\zeta} \\ x_{\zeta\zeta} &= \frac{\partial^2 x}{\partial \zeta^2} = -\frac{1}{\zeta^2} \\ (d\zeta)^2 &= \zeta(t)^2 \sum_{i=1}^{d} \lambda_i(t)^2 dt \end{aligned}$$

Inserting these expressions in dx yields

$$dx = -f(t,t)dt - \frac{1}{2} \sum_{i=1}^{d} \lambda_i(t)^2 dt + \sum_{i=1}^{d} \lambda_i(t) dz_i(t) \tag{A.4}$$

Since

$$x(T) = x(t) + \int_t^T dx(u) du$$

it follows that

$$\begin{aligned} x(T) &= x(t) + \int_t^T -f(u,u)du - \int_t^T \frac{1}{2} \sum_{i=1}^{d} \lambda_i(u)^2 du + \int_t^T \sum_{i=1}^{d} \lambda_i(u) dz_i(u) \\ &= x(t) + \int_t^T -f(u,u)du - \sum_{i=1}^{d} \int_t^T \frac{1}{2} \lambda_i(u)^2 du + \sum_{i=1}^{d} \int_t^T \lambda_i(u) dz_i(u) \end{aligned}$$

We insert x and apply the exponential function to both sides of the equation. Hence we obtain

$$\zeta(T) = \zeta(t) \exp\left(\int_t^T -f(u,u)du + \sum_{i=1}^d \int_t^T \lambda_i(u)dz_i(u) - \sum_{i=1}^d \int_t^T \frac{1}{2}\lambda_i(u)^2 du \right)$$

A.3 HJM Drift Condition

The dynamics of $Y(t,T) = \zeta(t)P(t,T)$ can be obtained by a straightforward application of Itô's lemma with $dP(t,T)$ from (3.5) and $d\zeta(t)$ from (3.11). According to Itô's lemma the dynamics of Y are

$$dY = Y_t dt + Y_\zeta d\zeta + Y_P dP + Y_{\zeta P}(d\zeta)(dP) + \frac{1}{2}Y_{\zeta\zeta}(d\zeta)^2 + \frac{1}{2}Y_{PP}(dP)^2$$

with

$$\begin{aligned}
Y_t &= 0 \\
Y_\zeta &= P \\
Y_P &= \zeta \\
Y_{\zeta P} &= 1 \\
Y_{\zeta\zeta} &= 0 \\
Y_{PP} &= 0 \\
(d\zeta)(dP) &= -P\zeta \sum_{i=1}^d \sigma_i(t,T)\lambda_i(t)dt
\end{aligned}$$

We insert these expressions in dY and obtain

$$\begin{aligned}
dY(t,T) =& P(t,T)\left[-f(t,t)\zeta(t)dt + \zeta(t)\sum_{i=1}^d \lambda_i(t)dz_i(t) \right] \\
&+ \zeta(t)\left[P(t,T)\mu(t,T) - P(t,T)\sum_{i=1}^d \sigma_i(t,T)dz_i(t) \right] \\
&- P(t,T)\zeta(t)\sum_{i=1}^d \sigma_i(t,T)\lambda_i(t)dt
\end{aligned}$$

Replacing $\zeta(t)P(t,T)$ with Y and simplification yields

$$\begin{aligned}
\frac{dY(t,T)}{Y(t,T)} =& \left(-f(t,t) + \mu(t,T) - \sum_{i=1}^d \sigma_i t, T\lambda_i(t) \right) dt \\
&+ \sum_{i=1}^d (\lambda_i(t) - \sigma_i(t,T))dz_i(t)
\end{aligned} \tag{A.5}$$

A.4 Special Case: Hull/White (1994)

The function $g(0,T)$ is defined as follows

$$
\begin{aligned}
&g(0,T) = \\
&-\frac{e^{-2T\kappa_r}\left(-1+2e^{T\kappa_r}\right)\kappa_\varepsilon\sigma_r^2}{\kappa_r\left(\kappa_r-\kappa_\varepsilon\right)^2}-\frac{e^{-2T\kappa_r}\kappa_\varepsilon\left(\left(-1+e^{T\kappa_r}\right)^2\kappa_\varepsilon-2e^{2T\kappa_r}\kappa_r\right)\sigma_r^2}{2\kappa_r^2\left(\kappa_r-\kappa_\varepsilon\right)^2}\\
&-\frac{e^{-2T\kappa_r}\left(-1+e^{T\kappa_r}\right)^2\sigma_r^2}{2\left(\kappa_r-\kappa_\varepsilon\right)^2}+\frac{e^{-T\kappa_r}\left(-1+e^{T\kappa_r}\right)\lambda_1\sigma_r}{\kappa_r}\\
&+\frac{\theta-e^{-T\kappa_r}\theta}{\kappa_r}+\frac{e^{-2T(\kappa_r+\kappa_\varepsilon)}\left(-1+e^{T\kappa_r}\right)}{\kappa_r^2\left(\kappa_r-\kappa_\varepsilon\right)\kappa_\varepsilon}\\
&\times\left(\varrho\left(e^{2T\kappa_\varepsilon}\left(-1+e^{T\kappa_r}\right)\kappa_\varepsilon-e^{T(\kappa_r+\kappa_\varepsilon)}\left(-1+e^{T\kappa_\varepsilon}\right)\kappa_r\right)\sigma_\varepsilon\sigma_r\right)\\
&-\frac{e^{-2T(\kappa_r+\kappa_\varepsilon)}\left(e^{T\kappa_r}\left(-1+e^{T\kappa_\varepsilon}\right)\kappa_r-e^{T\kappa_\varepsilon}\left(-1+e^{T\kappa_r}\right)\kappa_\varepsilon\right)^2\sigma_\varepsilon^2}{2\kappa_r^2\left(\kappa_r-\kappa_\varepsilon\right)^2\kappa_\varepsilon^2}\\
&+\frac{e^{-T(\kappa_r+\kappa_\varepsilon)}\varrho\left(e^{T\kappa_r}\left(-1+e^{T\kappa_\varepsilon}\right)\kappa_r-e^{T\kappa_\varepsilon}\left(-1+e^{T\kappa_r}\right)\kappa_\varepsilon\right)\lambda_1\sigma_\varepsilon}{\kappa_r\left(\kappa_r-\kappa_\varepsilon\right)\kappa_\varepsilon}\\
&+\frac{e^{-2T(\kappa_r+\kappa_\varepsilon)}\sqrt{1-\varrho^2}}{\kappa_r\left(\kappa_r-\kappa_\varepsilon\right)\kappa_\varepsilon}\\
&\times\left(e^{T(2\kappa_r+\kappa_\varepsilon)}\left(-1+e^{T\kappa_\varepsilon}\right)\kappa_r-e^{T(\kappa_r+2\kappa_\varepsilon)}\left(-1+e^{T\kappa_r}\right)\kappa_\varepsilon\right)\lambda_2\sigma_\varepsilon
\end{aligned}
\tag{A.6}
$$

B

Dynamic Bond Portfolio Optimization

The calculation of the expected values necessary for the derivation of the HJB equation is shown in detail in this appendix.

$$
\begin{aligned}
E(\Delta x \Delta x') &= (\alpha \Delta t + \beta \Delta z)(\alpha \Delta t + \beta \Delta z)' \\
&= (\alpha \Delta t + \beta \Delta z)(\alpha' \Delta t + \Delta z' \beta') \\
&= \beta \Delta z \Delta z' \beta' \\
&= \beta \underline{I} \Delta t \beta' \\
&= \beta \beta' \Delta t \\
E(\Delta x \Delta W) &= (\alpha \Delta t + \beta \Delta z)((r + w' \sigma \lambda) W \Delta t) - w' \sigma W \Delta z) \\
&= (\beta \Delta z)(w' \sigma \Delta z)(-W) \\
&= (\beta \Delta z)(\Delta z' \sigma' w)(-W) \\
&= \beta \underline{I} \Delta t \sigma' w(-W) \\
&= -\beta \sigma' w W \\
E(\Delta W^2) &= (w' \sigma \Delta z(-W))(w' \sigma \Delta z(-W)) \\
&= (w' \sigma \Delta z(-W))(\Delta z' \sigma' w(-W)) \\
&= w' \sigma \Delta z \Delta z' \sigma' w W^2 \\
&= w' \sigma \underline{I} \Delta t \sigma' w W^2 \\
&= w' \sigma \sigma' w W^2 \Delta t
\end{aligned}
$$

C

Dynamic Bond Portfolio Optimization

C.1 Vasicek (1977)

The function $g(t)$ is implicitly defined by the following equation

$$\begin{aligned}\ln(g(t)) =& \frac{e^{-2T\kappa}\left(e^{2t\kappa} + 3e^{2T\kappa} - 4e^{(t+T)\kappa}\right)\sigma^2\gamma^2}{4(\gamma-1)\kappa^3} \\ & - \frac{e^{-2T\kappa}\sigma\left(e^{2T\kappa}(2\lambda + (T-t)\sigma) - 2e^{(t+T)\kappa}\lambda\right)\gamma^2}{2(\gamma-1)\kappa^2} + (T-t)\theta\gamma \\ & + \frac{(t-T)\lambda(\kappa\lambda - 2\gamma\sigma)\gamma}{2(\gamma-1)\kappa} + \frac{\left(-1 + e^{(t-T)\kappa}\right)\theta\gamma}{\kappa}\end{aligned} \tag{C.1}$$

C.2 Hull/White (1994)

The function $g(t)$ is implicitly defined by the following equation

$$
\begin{aligned}
\ln(g(t)) = \\
&\frac{T\theta\kappa_\epsilon^2\gamma^2}{(\gamma-1)\kappa_r\left(\kappa_r-\kappa_\epsilon\right)^2}-\frac{\theta\kappa_\epsilon^2\gamma^2}{(\gamma-1)\kappa_r^2\left(\kappa_r-\kappa_\epsilon\right)^2}\\
&+\frac{\left(2T\kappa_r-3\right)\left(2\kappa_r-\kappa_\epsilon\right)\kappa_\epsilon\sigma_r^2\gamma^2}{4(\gamma-1)\kappa_r^3\left(\kappa_r-\kappa_\epsilon\right)^2}+\frac{\left(3-2T\kappa_r\right)\sigma_r^2\gamma^2}{4(\gamma-1)\kappa_r\left(\kappa_r-\kappa_\epsilon\right)^2}\\
&+\frac{\left(\left(3-2T\kappa_\epsilon\right)\kappa_r^2+\kappa_\epsilon\left(5-2T\kappa_\epsilon\right)\kappa_r+3\kappa_\epsilon^2\right)\sigma_\epsilon^2\gamma^2}{4(\gamma-1)\kappa_r^3\kappa_\epsilon^3\left(\kappa_r+\kappa_\epsilon\right)}+\frac{2\theta\kappa_\epsilon\gamma^2}{(\gamma-1)\kappa_r\left(\kappa_r-\kappa_\epsilon\right)^2}\\
&+\frac{\left(\kappa_r\left(T\kappa_\epsilon-1\right)-\kappa_\epsilon\right)\left(\varrho\lambda_1+\sqrt{1-\varrho^2}\lambda_2\right)\sigma_\epsilon\gamma^2}{(\gamma-1)\kappa_r^2\kappa_\epsilon^2}\\
&+\frac{\varrho\left(\left(2-2T\kappa_\epsilon\right)\kappa_r^2-2\kappa_\epsilon\left(T\kappa_\epsilon-2\right)\kappa_r+3\kappa_\epsilon^2\right)\sigma_r\sigma_\epsilon\gamma^2}{2(\gamma-1)\kappa_r^3\kappa_\epsilon^2\left(\kappa_r+\kappa_\epsilon\right)}\\
&-\frac{T\theta\kappa_\epsilon^2\gamma}{(\gamma-1)\kappa_r\left(\kappa_r-\kappa_\epsilon\right)^2}+\frac{\theta\kappa_\epsilon^2\gamma}{(\gamma-1)\kappa_r^2\left(\kappa_r-\kappa_\epsilon\right)^2}-\frac{2\theta\left(T(\gamma-1)\kappa_r+1\right)\kappa_\epsilon\gamma}{(\gamma-1)\kappa_r\left(\kappa_r-\kappa_\epsilon\right)^2}\\
&+\frac{\left(-T\left(\lambda_1^2+\lambda_2^2\right)\kappa_r^2+2T\gamma\lambda_1\sigma_r\kappa_r-2\gamma\lambda_1\sigma_r\right)\gamma}{2(\gamma-1)\kappa_r^2}-\frac{\theta\gamma}{\left(\kappa_r-\kappa_\epsilon\right)^2}\\
&+\frac{T\theta\kappa_r\gamma}{\left(\kappa_r-\kappa_\epsilon\right)^2}
\end{aligned}
\tag{C.2}
$$

C.3 International Bond Portfolio Selection

The function $g(t)$ is implicitly defined by the following equation

$$
\begin{aligned}
\ln(g(t)) = \\
&\frac{e^{-2T\kappa_1}\left(e^{2t\kappa_1}+3e^{2T\kappa_1}-4e^{(t+T)\kappa_1}\right)\sigma_{r1}^2\gamma^2}{4(\gamma-1)\kappa_1^3}+\frac{(T-t)\lambda_1\sigma_{r1}\gamma^2}{(\gamma-1)\kappa_1}-\frac{\lambda_1\sigma_{r1}\gamma^2}{(\gamma-1)\kappa_1^2}\\
&+\frac{e^{-2T\kappa_1}\sigma_{r1}\left(2e^{(t+T)\kappa_1}\lambda_1-e^{2T\kappa_1}(T-t)\sigma_{r1}\right)\gamma^2}{2(\gamma-1)\kappa_1^2}+(T-t)\theta_1\gamma\\
&+\frac{e^{-T\kappa_1}\left(2\left(e^{t\kappa_1}-e^{T\kappa_1}\right)(\gamma-1)\theta_1-e^{T\kappa_1}(T-t)\kappa_1\left(\lambda_1^2+\lambda_2^2\right)\right)\gamma}{2(\gamma-1)\kappa_1}
\end{aligned}
\tag{C.3}
$$

References

1. Agca, S. (2002) The Performance of Alternative Interest Rate Risk Measures and Immunization Strategies Under a Heath-Jarrow-Morton Framework. Ph. D thesis, Virginia Polytechnic Institute and State University
2. Amin, K. I., Jarrow, R. A. (1991) Pricing foreign currency options under stochastic interest rates. Journal of International Money and Finance 10, pp. 310–329
3. Anderson, N. et al. (1996) Estimating and Interpreting The Yield Curve. Chicester: John Wiley & Sons Ltd.
4. Bank, P., Baum, D. (2004) Hedging and Portfolio Optimization in Financial Markets with a Large Trader. Mathematical Finance 14 (1), pp. 1–18
5. Baz, J., Chacko, G. (2004) Financial Derivatives. Pricing, Applicatons, and Mathematics. Cambridge: Cambridge University Press
6. Bieri, D. S., Chincarini, L. B. (2005) Riding the Yield Curve: A Variety of Strategies. Journal of Fixed Income 15 (2), pp. 6–35
7. Bierwag, G. O., Fooladi, I., Roberts, G. S. (1993) Designing an immunized portfolio: Is M-squared the key? Journal of Banking and Finance 17 (6), pp. 1147–1170
8. Björk, T. (1998) Arbitrage Theory in Continuous Time. Oxford: Oxford University Press
9. Black, F., Karasinksi, P. (1991) Bond and Option Pricing when Short Rates are Lognormal. Financial Analysts Journal 47 (4), pp. 52–59
10. Bradley, S. P., Crane, D. B. (1972) A Dynamic Model for Bond Portfolio Management. Management Science 19 (2), pp. 139–151
11. Branger, N., Schlag, C. (2004) Zinsderivate. Modelle und Bewertung. Berlin et. al.: Springer Verlag
12. Brennan, M. J., Schwartz, E. S. (1980) Conditional predictions of bond prices and returns. Journal of Finance 35, pp. 405–417
13. Brigo, D., Mercurio, F. (2001) Interest Rate Models: Theory and Practice. Berlin et. al.: Springer
14. Bronstein, I. et al. (1999) Taschenbuch der Mathematik. 4th edition. Frankfurt am Main: Verlag Harri Deutsch

15. Brown, P. J. (1994) Constructing And Calculating Bond Indices. Cambridge: Probus Publishing Company
16. Cairns, A. J. G. (2004) Interest Rate Models. Princeton: Princeton University Press
17. Caks, J. (1977) The Coupon Effect On Yield To Maturity. The Journal of Finance 32 (1), pp. 103–115
18. Chan, L. K. C. et al. (1992) An Empricial Comparison of Alternative Models of the Short-Term Interest Rate. The Journal of Finance 47 (3), pp. 1209–1227
19. Cheng, P. L. (1962) Optimum Bond Portfolio Selection. Management Science 8 (4), pp. 490–499
20. Choudhry, M. (2004) Analysing and Interpreting the Yield Curve. Chichester: John Wiley & Sons
21. Cochrane, J. H. (2005) Asset Pricing. revised edition. Princeton: Princeton University Press
22. Cox, J., Huang, C. F. (1989) Optimal consumption and portfolio policies when asset prices follow a diffusion process. Journal of Economic Theory 49, pp. 33–83
23. Cox, J. C., Ingersoll, J. E., Ross, S. A. (1981) A Re-examination of Traditional Hypotheses about the Term Structure of Interest Rates. The Journal of Finance 36 (4), pp. 769–799
24. Cox, J. C., Ingersoll, J. E., Ross, S. A. (1985) A Theory Of The Term Structure Of Interest Rates. Econometrica 53 (2), pp. 385–407
25. Culbertson, J. M. (1957) The Term Structure of Interest Rates. Quarterly Journal of Economics 82, pp. 489–504
26. Cuoco, D., Cvitanic, J. (1998) Optimal consumption choices for a "large" investor. Journal of Economic Dynamics and Control 22 (3), pp. 401–436
27. Cvitanic, J., Karatzas, I. (1992) Convex Duality in Constrained Portfolio Optimization. Annals of Applied Probability 2 (4), pp. 767–818
28. Cvitanic, J., Zapatero, F. (2004) Introduction to the Economics and Mathematics of Financial Markets. Cambridge: MIT Press
29. Dothan, L. U. (1978) On the Term Structure of Interest Rates. Journal of Financial Economics 6, pp. 59–69
30. Duffie, D. (1996) Dynamic Asset Pricing Theory. 2nd edition. Princeton: Princeton University Press
31. Duffie, D., Kan, R. (1996) A Yield-Factor Model of Interest Rates. Mathematical Finance 6 (4), pp. 379–406
32. Elton, E. J. et al. (2003) Modern Portfolio Theory and Investment Analysis. Chichester: John Wiley & Sons
33. Fabozzi, F. J. (2000) The Handbook of Fixed Income Securities. 6th edition. Chichester: John Wiley & Sons
34. Fabozzi, F. J. (2001) Bond Portfolio Management. New Hope: Frank J. Fabozzi Associates

35. Fabozzi, F. J. (2004) Bond Markets, Analysis, and Strategies. 6th edition. New York: Prentice Hall
36. Fabozzi, F. J., Fong, G. (1994) Advanced Fixed Income Portfolio Management. Chicago: Probus Publishing Company
37. Flesaker, B., Hughston, L. P. (2000) International Models for Interest Rates and Foreign Exchange. In Hughston, L. P., editor The New Interest Rate Models. Recent Developments in the Theory and Application of Yield Curve Dynamics. London, UK: RiskBooks
38. Focardi, S. M., Fabozzi, F. J. (2004) The Mathematics of Financial Modeling and Investment Management. Chichester: John Wiley & Sons
39. Fong, H. G., Vasicek, O. (1984) A Risk Minimizing Strategy for Multiple Liability Immunization. Journal of Finance 39 (12), pp. 1541–1546
40. Garbade, K. D. (1996) Fixed Income Analytics. Cambridge: MIT Press
41. Golub, B. W., Tilman, L. M. (2000) Risk Management. Approaches for Fixed Income Markets. New York: John Wiley & Sons
42. Granito, M. R. (1987) The problem with bond index funds. The Journal of Portfolio Management (Summer), pp. 41–47
43. Grinold, R. C., Kahn, R. N. (2000) Active Portfolio Management. New York: McGraw Hill
44. Harrison, J. M., Pliska, S. R. (1981) Martingales and Stochastic Integrals in the Theory of Continuous Trading. Stochastic Processes and Their Applications 11, pp. 215–260
45. Harrison, M. J., Kreps, D. M. (1979) Martingales and Arbitrage in Multiperiod Securities Markets. Journal of Economic Theory 20, pp. 381–408
46. Heath, D., Jarrow, R., Morton, A. (1992) Bond Pricing And The Term Structure Of Interest Rates: A New Methodology For Contingent Claims Valuation. Econometrica 60 (1), pp. 77–105
47. Hicks, J. H. (1939) Value and Capital. Oxford: Clarendon Press
48. Ho, T. S., Lee, S.-B. (1986) Term Structure Movements and Pricing Interest Rate Contingent Claims. The Journal of Finance 41 (5), pp. 1011–1029
49. Holden, C. W. (2005) Excel Modelling in Investments. New York: Pearson Prentice Hall
50. Hull, J., White, A. (1990) Pricing Interest-Rate-Derivative Securities. The Review of Financial Studies 3 (4), pp. 573–592
51. Hull, J., White, A. (1994) Numerical Procedures for Implementing Term Structure Models II: Two-Factor Models. The Journal of Derivatives 2 (2), pp. 37–48
52. Hull, J., White, A. (1996) Hull-White on Derivatives. A Compilation of Articles. London: Risk Books
53. Hull, J. C. (2005) Options, Futures and Other Derivatives. 6th edition. New York: Prentice Hall
54. Ilmanen, A. (1995) Understanding the Yield Curve: Part 1 - Overview of Forward Rate Analysis. Salomon Brothers – Technical report

55. Ingersoll, J. E. (1987) Theory of Financial Decision Making. Savage: Rowman & Litterfield

56. James, W., Stein, C. (1961) Estimation with quadratic loss. In Proceedings of the Fourth Berkeley Symposium on Mathematical Statistics and Probability 1. University of California Press, pp. 361–379

57. Jin, Y., Glasserman, P. (2001) Equilibrium Positive Interest Rates: A Unified View. The Review of Financial Studies 14, pp. 187–214

58. Jones, F. J. (1991) Yield Curve Strategies. The Journal of Fixed Income 1 (2), pp. 43–51

59. Jorion, P. (1986) Bayes-Stein Estimation for Portfolio Analysis. Journal of Financial and Quantitative Analysis 21 (3), pp. 279–292

60. Kamien, M. I., Schwartz, N. L. (1981) Dynamic Optimization: The Calculus of Variations and Optimal Control in Economics and Management. Amsterdam: Elsevier Science Publishing

61. Korn, O., Koziol, C. (2006) Bond Portfolio Optimization: A Risk-Return Approach. Journal of Fixed Income 15 (4), pp. 48–60

62. Korn, R. (1997) Optimal Portfolios. Singapore et. al.: World Scientific

63. Korn, R., Kraft, H. (2002) A Stochastic Control Approach To Portfolio Problems With Stochastic Interest Rates. SIAM Journal on Control and Optimization 40 (4), pp. 1250–1269

64. Kraft, H. (2004) Optimal Portfolio with Stochastic Interest Rates and Defaultable Assets. Berlin et. al.: Springer

65. Krgin, D. (2002) Handbook of Global Fixed Income Calculations. New York: John Wiley & Sons, Inc.

66. Kushner, H. J., Dupuis, P. (2000) Numerical Methods for Stochastic Control Problems in Continuous Time. Berlin et. al.: Springer Verlag

67. Langetieg, T. C. (1980) A Multivariate Model of the Term Structure. Journal of Finance 35, pp. 71–97

68. Ledoit, O., Wolfe, M. (2004) Honey, I Shrunk the Sample Covariance Matrix. The Journal of Portfolio Management 31 (1), pp. 110–119

69. Lipton, A. (2001) Mathematical Methods for Foreign Exchange. Singapore et. al.: World Scientific

70. Litterman, B. (2003) Modern Investment Management. Chichester: John Wiley & Sons

71. Litterman, R., Scheinkman, J. (1991) Common Factors Affecting Bond Returns. The Journal of Fixed Income 1 (1), pp. 54–61

72. Lutz, F. A. (1940) The Structure Of Interest Rates. Quarterly Journal of Economics 55 (1), pp. 36–63

73. Macaulay, F. R. (1938) Some Theoretical Problems Suggested by the Movements of Interest Rates, Bond Yields, and Stock Prices in the United States since 1856. New York: Columbia University Press

74. Malliaris, A. G., Brock, W. A. (1982) Stochastic Methods in Economics and Finance. Amsterdam: North Holland

75. Mann, S. V., Ramanlal, P. (1997) The Relative Performance of Yield Curve Strategies. Journal of Portfolio Management 23, pp. 64–70

76. Markowitz, H. (1956) The Optimization of a Quadratic Function Subject to Linear Constraints. Naval Research Logistics Quarterly 3, pp. 111–133
77. Markowitz, H. (1959) Portfolio Selection. Efficient Diversification of Investments. New Haven et. al.: Yale University Press
78. Markowitz, H. M. (1952) Portfolio Selection. The Journal of Finance 7 (1), pp. 77–91
79. Markowitz, H. M., Todd, G. P. (2000) Mean-Variance Analysis in Portfolio Choice and Capital Markets. Chichester: John Wiley & Sons
80. Martellini, L., Priaulet, P., Priaulet, S. (2003) Fixed-Income Securities. Chichester: John Wiley & Sons, Ltd
81. McCulloch, J. H. (1971) Measuring the Term Structure of Interest Rates. Journal of Business 44 (1), pp. 19–31
82. Mertens, D. (2006) Portfolio-Optimierung nach Markowitz. 2nd edition. Frankfurt am Main: Bankakademie Verlag
83. Merton, R. C. (1969) Lifetime Portfolio Selection Under Uncertainty: The Continous-Time Case. The Review of Economics and Statistics 51, pp. 247–257
84. Merton, R. C. (1971) Optimum Consumption and Portfolio Rules in a Continuous-Time Model. Journal of Economic Theory 3, pp. 373–413
85. Merton, R. C. (1992) Continous-Time Finance. Oxford: Blackwell
86. Modigliani, F., Sutch, R. (1966) Innovations in Interest Rate Policy. American Economic Review 56 (1/2), pp. 178–197
87. Mossavar-Rahmani, S. (1991) Bond Index Funds. Chicago: Probus Publishing Company
88. Mossin, J. (1968) Optimal Multiperiod Portfolio Policies. Journal of Business 41 (2), pp. 215–229
89. Munk, C. (2004a) Dynamic Asset Allocation. Lecture Notes. ⟨URL: `http://www.sam.sdu.dk/~cmu/noter/dynassal.pdf`⟩
90. Munk, C. (2004b) Fixed Income Analysis: Securities, Pricing, and Risk Management. ⟨URL: `http://www.sam.sdu.dk/~cmu/noter/FIAjan05.pdf`⟩
91. Munk, C., Sørensen, C. (2004) Optimal consumption and investment strategies with stochastic interest rates. Journal of Banking & Finance 28, pp. 1987–2013
92. Musiela, Rutkowski (1997) Martingale Methods in Financial Modelling. Berlin et. al.: Springer Verlag
93. Nelson, C. R., Siegel, A. F. (1987) Parsimonious Modeling of Yield Curves. Journal of Business 60 (4), pp. 473–489
94. Nietert, B. (1996) Dynamische Portfolio-Selektion. Karlsruhe: VVW
95. Oksendal, B. (1992) Stochastic Differential Equations. 6th edition. Berlin et. al.: Springer Verlag
96. Pliska, S. R. (1986) A stochastic calculus model of continuous trading: Optimal portfolios. Mathematics of Operations Research 1, pp. 100–124
97. Rebonato, R. (1998) Interest-Rate Option Models. 2nd edition. Chichester: John Wiley & Sons

98. Redington, F. M. (1952) Review of the Principles of Life-Office Valuations. Journal of the Institute of Actuaries 78 (3), pp. 286–315

99. Rinne, H. (1997) Taschenbuch der Statistik. 2nd edition. Frankfurt am Main: Verlag Harri Deutsch

100. Rudolf, M. (1994) Algorithms for Portfolio Optimization and Portfolio Insurance. Ph. D thesis, Hochschule St. Gallen für Wirtschafts-, Rechts- und Sozialwissenschaften

101. Scherer, B. (2002) Portfolio Construction and Risk Budgeting. London: Risk Books

102. Shiller, R. J. (1990) The Term Structure of Interest Rates. Elsevier Science Publishers B.V., Handbook of Monetary Economics, Volume I, pp. 627–722

103. Sørensen, C. (1999) Dynamic Asset Allocation and Fixed Income Management. Journal of Financial and Quantitative Analysis 34 (4), pp. 513–531

104. Stigum, M., Robinson, F. L. (1996) Money Market and Bond Market Calculations. Rev. edition. Chicago: Irwin

105. Svensson, L. E. O. (1994) Estimating And Interpreting Forward Interest Rates: Sweden 1992-1994. NBER Working Paper No. 4871

106. Svoboda, S. (2004) Interest rate modelling. Basingstoke: Palgrave Macmillan

107. Tuckman, B. (2002) Fixed Income Securities. Chichester: John Wiley & Sons

108. Vasicek, O. (1977) An Equilibrium Characterization Of The Term Structure. Journal of Financial Economics 5 (8), pp. 177–188

109. Wilhelm, J. (1992) Fristigkeitsstruktur und Zinsänderungsrisiko - Vorüberlegungen zu einer Markowitz-Theorie des Bond-Portfolio-Managements. Zeitschrift für betriebswirtschaftliche Forschung 44 (3), pp. 209–246

110. Wilhelm, J. (1995) Zinsstruktur. Stuttgart: Schäffer-Poeschel, Handwörterbuch des Bank- und Finanzwesens, pp. 2052–2059

111. Wilhelm, J. (1996) Stochastic Discounting and the Term Structure of Interest Rates in Discrete Time. Working Paper 1/1996. ⟨URL: http://www.wiwi.uni-passau.de/fileadmin/dokumente/lehrstuehle/wilhelm/pdf/Stochastic_Discounting.pdf⟩

112. Wilhelm, J. (1999) A fresh view on the Ho-Lee model of the term structure from a stochastic discounting perspective. OR Spektrum 24, pp. 9–34

113. Wilhelm, J. (2005) The Gaussian Interest Rate Model and Related Issues. Working Paper AP Aug 2005. ⟨URL: http://www.wiwi.uni-passau.de/fileadmin/dokumente/lehrstuehle/wilhelm/pdf/APBSchHoLee_3.pdf⟩

114. Willner, R. (1996) A New Tool for Portfolio Managers: Level, Slope, and Curvature Durations. The Journal of Fixed Income 6 (1), pp. 48–59

115. Xu, G.-L., Shreve, S. E. (1992a) A Duality Method for Optimal Consumption and Investment Under Short-Selling Prohibition. I. General Market Coefficients. Annals of Applied Probability 2 (1), pp. 87–112
116. Xu, G.-L., Shreve, S. E. (1992b) A Duality Method for Optimal Consumption and Investment Under Short-Selling Prohibition: II. Constant Market Coefficients. Annals of Applied Probability 2 (2), pp. 314–328
117. Zimmerer, T. (2003) Der Roll Down im Bondmanagement: Theoretisches Phänomen und praktische Implementierung. FinanzBetrieb 4, pp. 243–252

List of Tables

List of Figures

Lecture Notes in Economics and Mathematical Systems

For information about Vols. 1–519
please contact your bookseller or Springer-Verlag

Vol. 520: B. Glaser, Efficiency versus Sustainability in Dynamic Decision Making. IX, 252 pages. 2002.

Vol. 521: R. Cowan, N. Jonard (Eds.), Heterogenous Agents, Interactions and Economic Performance. XIV, 339 pages. 2003.

Vol. 522: C. Neff, Corporate Finance, Innovation, and Strategic Competition. IX, 218 pages. 2003.

Vol. 523: W.-B. Zhang, A Theory of Interregional Dynamics. XI, 231 pages. 2003.

Vol. 524: M. Frölich, Programme Evaluation and Treatment Choise. VIII, 191 pages. 2003.

Vol. 525: S. Spinler, Capacity Reservation for Capital-Intensive Technologies. XVI, 139 pages. 2003.

Vol. 526: C. F. Daganzo, A Theory of Supply Chains. VIII, 123 pages. 2003.

Vol. 527: C. E. Metz, Information Dissemination in Currency Crises. XI, 231 pages. 2003.

Vol. 528: R. Stolletz, Performance Analysis and Optimization of Inbound Call Centers. X, 219 pages. 2003.

Vol. 529: W. Krabs, S. W. Pickl, Analysis, Controllability and Optimization of Time-Discrete Systems and Dynamical Games. XII, 187 pages. 2003.

Vol. 530: R. Wapler, Unemployment, Market Structure and Growth. XXVII, 207 pages. 2003.

Vol. 531: M. Gallegati, A. Kirman, M. Marsili (Eds.), The Complex Dynamics of Economic Interaction. XV, 402 pages, 2004.

Vol. 532: K. Marti, Y. Ermoliev, G. Pflug (Eds.), Dynamic Stochastic Optimization. VIII, 336 pages. 2004.

Vol. 533: G. Dudek, Collaborative Planning in Supply Chains. X, 234 pages. 2004.

Vol. 534: M. Runkel, Environmental and Resource Policy for Consumer Durables. X, 197 pages. 2004.

Vol. 535: X. Gandibleux, M. Sevaux, K. Sörensen, V. T'kindt (Eds.), Metaheuristics for Multiobjective Optimisation. IX, 249 pages. 2004.

Vol. 536: R. Brüggemann, Model Reduction Methods for Vector Autoregressive Processes. X, 218 pages. 2004.

Vol. 537: A. Esser, Pricing in (In)Complete Markets. XI, 122 pages, 2004.

Vol. 538: S. Kokot, The Econometrics of Sequential Trade Models. XI, 193 pages. 2004.

Vol. 539: N. Hautsch, Modelling Irregularly Spaced Financial Data. XII, 291 pages. 2004.

Vol. 540: H. Kraft, Optimal Portfolios with Stochastic Interest Rates and Defaultable Assets. X, 173 pages. 2004.

Vol. 541: G.-y. Chen, X. Huang, X. Yang, Vector Optimization. X, 306 pages. 2005.

Vol. 542: J. Lingens, Union Wage Bargaining and Economic Growth. XIII, 199 pages. 2004.

Vol. 543: C. Benkert, Default Risk in Bond and Credit Derivatives Markets. IX, 135 pages. 2004.

Vol. 544: B. Fleischmann, A. Klose, Distribution Logistics. X, 284 pages. 2004.

Vol. 545: R. Hafner, Stochastic Implied Volatility. XI, 229 pages. 2004.

Vol. 546: D. Quadt, Lot-Sizing and Scheduling for Flexible Flow Lines. XVIII, 227 pages. 2004.

Vol. 547: M. Wildi, Signal Extraction. XI, 279 pages. 2005.

Vol. 548: D. Kuhn, Generalized Bounds for Convex Multistage Stochastic Programs. XI, 190 pages. 2005.

Vol. 549: G. N. Krieg, Kanban-Controlled Manufacturing Systems. IX, 236 pages. 2005.

Vol. 550: T. Lux, S. Reitz, E. Samanidou, Nonlinear Dynamics and Heterogeneous Interacting Agents. XIII, 327 pages. 2005.

Vol. 551: J. Leskow, M. Puchet Anyul, L. F. Punzo, New Tools of Economic Dynamics. XIX, 392 pages. 2005.

Vol. 552: C. Suerie, Time Continuity in Discrete Time Models. XVIII, 229 pages. 2005.

Vol. 553: B. Mönch, Strategic Trading in Illiquid Markets. XIII, 116 pages. 2005.

Vol. 554: R. Foellmi, Consumption Structure and Macro-economics. IX, 152 pages. 2005.

Vol. 555: J. Wenzelburger, Learning in Economic Systems with Expectations Feedback (planned) 2005.

Vol. 556: R. Branzei, D. Dimitrov, S. Tijs, Models in Cooperative Game Theory. VIII, 135 pages. 2005.

Vol. 557: S. Barbaro, Equity and Efficiency Considerations of Public Higer Education. XII, 128 pages. 2005.

Vol. 558: M. Faliva, M. G. Zoia, Topics in Dynamic Model Analysis. X, 144 pages. 2005.

Vol. 559: M. Schulmerich, Real Options Valuation. XVI, 357 pages. 2005.

Vol. 560: A. von Schemde, Index and Stability in Bimatrix Games. X, 151 pages. 2005.

Vol. 561: H. Bobzin, Principles of Network Economics. XX, 390 pages. 2006.

Vol. 562: T. Langenberg, Standardization and Expectations. IX, 132 pages. 2006.

Vol. 563: A. Seeger (Ed.), Recent Advances in Optimization. XI, 455 pages. 2006.

Vol. 564: P. Mathieu, B. Beaufils, O. Brandouy (Eds.), Artificial Economics. XIII, 237 pages. 2005.

Vol. 565: W. Lemke, Term Structure Modeling and Estimation in a State Space Framework. IX, 224 pages. 2006.

Vol. 566: M. Genser, A Structural Framework for the Pricing of Corporate Securities. XIX, 176 pages. 2006.

Vol. 567: A. Namatame, T. Kaizouji, Y. Aruga (Eds.), The Complex Networks of Economic Interactions. XI, 343 pages. 2006.

Vol. 568: M. Caliendo, Microeconometric Evaluation of Labour Market Policies. XVII, 258 pages. 2006.

Vol. 569: L. Neubecker, Strategic Competition in Oligopolies with Fluctuating Demand. IX, 233 pages. 2006.

Vol. 570: J. Woo, The Political Economy of Fiscal Policy. X, 169 pages. 2006.

Vol. 571: T. Herwig, Market-Conform Valuation of Options. VIII, 104 pages. 2006.

Vol. 572: M. F. Jäkel, Pensionomics. XII, 316 pages. 2006

Vol. 573: J. Emami Namini, International Trade and Multinational Activity. X, 159 pages, 2006.

Vol. 574: R. Kleber, Dynamic Inventory Management in Reverse Logistics. XII, 181 pages, 2006.

Vol. 575: R. Hellermann, Capacity Options for Revenue Management. XV, 199 pages, 2006.

Vol. 576: J. Zajac, Economics Dynamics, Information and Equilibnum. X, 284 pages, 2006.

Vol. 577: K. Rudolph, Bargaining Power Effects in Financial Contracting. XVIII, 330 pages, 2006.

Vol. 578: J. Kühn, Optimal Risk-Return Trade-Offs of Commercial Banks. IX, 149 pages, 2006.

Vol. 579: D. Sondermann, Introduction to Stochastic Calculus for Finance. X, 136 pages, 2006.

Vol. 580: S. Seifert, Posted Price Offers in Internet Auction Markets. IX, 186 pages, 2006.

Vol. 581: K. Marti; Y. Ermoliev; M. Makowsk; G. Pflug (Eds.), Coping with Uncertainty. XIII, 330 pages, 2006.

Vol. 582: J. Andritzky, Sovereign Default Risks Valuation: Implications of Debt Crises and Bond Restructurings. VIII, 251 pages, 2006.

Vol. 583: I.V. Konnov, D.T. Luc, A.M. Rubinov† (Eds.), Generalized Convexity and Related Topics. IX, 469 pages, 2006.

Vol. 584: C. Bruun, Adances in Artificial Economics: The Economy as a Complex Dynamic System. XVI, 296 pages, 2006.

Vol. 585: R. Pope, J. Leitner, U. Leopold-Wildburger, The Knowledge Ahead Approach to Risk. XVI, 218 pages, 2007 (planned).

Vol. 586: B.Lebreton, Strategic Closed-Loop Supply Chain Management. X, 150 pages, 2007 (planned).

Vol. 587: P. N. Baecker, Real Options and Intellectual Property: Capital Budgeting Under Imperfect Patent Protection. X, 276 pages , 2007.

Vol. 588: D. Grundel, R. Murphey, P. Panos , O. Prokopyev (Eds.), Cooperative Systems: Control and Optimization. IX, 401 pages , 2007.

Vol. 589: M. Schwind, Dynamic Pricing and Automated Resource Allocation for Information Services: Reinforcement Learning and Combinatorial Auctions. XII, 293 pages , 2007.

Vol. 590: S. H. Oda, Developments on Experimental Economics: New Approaches to Solving Real-World Problems. XVI, 262 pages, 2007.

Vol. 591: M. Lehmann-Waffenschmidt, Economic Evolution and Equilibrium: Bridging the Gap. VIII, 272 pages, 2007.

Vol. 592: A. C.-L. Chian, Complex Systems Approach to Economic Dynamics. X, 95 pages, 2007.

Vol. 593: J. Rubart, The Employment Effects of Technological Change: Heterogenous Labor, Wage Inequality and Unemployment. XII, 209 pages, 2007

Vol. 594: R. Hübner, Strategic Supply Chain Management in Process Industries: An Application to Specialty Chemicals Production Network Design. XII, 243 pages, 2007

Vol. 595: H. Gimpel, Preferences in Negotiations: The Attachment Effect. XIV, 268 pages, 2007

Vol. 596: M. Müller-Bungart, Revenue Management with Flexible Products: Models and Methods for the Broadcasting Industry. XXI, 297 pages, 2007

Vol. 597: C. Barz, Risk-Averse Capacity Control in Revenue Management. XIV, 163 pages, 2007

Vol. 598: A. Ule, Partner Choice and Cooperation in Networks: Theory and Experimental Evidence. Approx. 200 pages, 2007

Vol. 599: A. Consiglio, Artificial Markets Modeling: Methods and Applications. XV, 277 pages, 2007

Vol. 600: M. Hickman, P. Mirchandani, S. Voss (Eds.): Computer-Aided Scheduling of Public Transport. Approx. 424 pages, 2007

Vol. 601: D. Radulescu, CGE Models and Capital Income Tax Reforms: The Case of a Dual Income Tax for Germany. XVI, 168 pages, 2007

Vol. 602: N. Ehrentreich, Agent-Based Modeling: The Santa Fe Institute Artificial Stock Market Model Revisited. XVI, 225 pages, 2007

Vol. 603: D. Briskorn, Sports Leagues Scheduling: Models, Combinatorial Properties, and Optimization Algorithms. XII, 164 pages, 2008

Vol. 604: D. Brown, F. Kubler, Computational Aspects of General Equilibrium Theory: Refutable Theories of Value. XII, 202 pages, 2008

Vol. 605: M. Puhle, Bond Portfolio Optimization. XIV, 137 pages, 2008

Vol. 606: S. von Widekind, Evolution of Non-Expected Utility Preferences. X, 130 pages, 2008

Vol. 607: M. Bouziane, Pricing Interest Rate Derivatives: A Fourier-Transform Based Approach. XII, 191 pages, 2008

Made in the USA
Monee, IL
13 September 2026

62975071R00087